T0325815

Nothing but the Tooth

Nothing but the Tooth

A Dental Odyssey

Barry K.B. Berkovitz

Emeritus Reader in Anatomy,
School of Biomedical Sciences,
King's College London,
London, UK

ELSEVIER

AMSTERDAM • BOSTON • HEIDELBERG • LONDON • NEW YORK • OXFORD
PARIS • SAN DIEGO • SAN FRANCISCO • SINGAPORE • SYDNEY • TOKYO

Elsevier
32 Jamestown Road, London NW1 7BY
225 Wyman Street, Waltham, MA 02451, USA

First edition 2013

British Library Cataloguing-in-Publication Data
A catalogue record for this book is available from the British Library

Library of Congress Cataloging-in-Publication Data
A catalog record for this book is available from the Library of Congress

ISBN: 978-0-12-397190-6

For information on all Elsevier publications
visit our website at store.elsevier.com

This book has been manufactured using Print On Demand technology. Each copy is produced to order and is limited to black ink. The online version of this book will show color figures where appropriate.

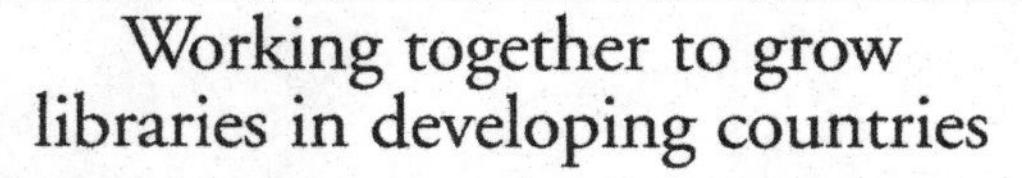

Dedication

To my namesake Kenneth Bradbury – killed in the defence of Teruel, Spain in 1938, during the Spanish Civil War, aged 17.

Barry Kenneth Bradbury Berkovitz

Contents

Acknowledgements

I am extremely grateful to the following for their help and constructive criticism:

Dr S.R. Berkovitz (Chapter 1), Dr R.P. Shellis (Chapters 2 and 9), Professor C.S.C. Liu (Chapter 5), Dr N. Sahara, (Chapter 8), Professor P.-I. Branemark (Chapter 10), Professor R. Palmer (Chapter 10), Professor M.C. Dean (Chapter 13), Ms M. Farrell (Chapters 12 and 16) and my wife, Sylvia, for her assistance throughout.

My special thanks to Mr M. Simon for his invaluable help with all the images and illustrations.

Barry K.B. Berkovitz

Preface

I retired in 2005 after spending over 40 years teaching and researching the structure and function of teeth. Over that time, I have amassed a collection of information that I thought might be of interest to a wider audience. I imagined that at some stage someone surely would ask me to give a talk on the subject. Having time on my hands, I decided to go ahead and prepare a non-clinical, illustrated, 45-min talk demonstrating how enjoyable the subject could be. The first thing I did was to write out a list of topics. However, when I looked at it, I realised that I was ignorant of much of the finer detail of the subject matter and would have to spend time mugging it up. Four years later, I finally finished preparing my talk, but as yet I have not been invited to deliver it. Therefore, I decided to publish it in the form of this book. If each reader finds at least one topic of interest, I will feel that my efforts have not been in vain.

Barry K.B. Berkovitz
London, June 2012

About the Author

Dr Barry K.B. Berkovitz is an internationally recognised teacher, examiner and research worker in the field of dentistry, with over 40 years' experience. This is his 14th book and he has written well over 120 major scientific articles. He qualified at the Royal Dental Hospital, London, and undertook postgraduate research at Royal Holloway College. He subsequently taught at the University of Bristol and King's College London. He is the Honorary Curator of the Odontological Collection at the Hunterian Museum of the Royal College of Surgeons of England and an Honorary Research Associate in the History of Dentistry Unit at the Dental Institute, King's College London.

1 The Jaws of the Piranha

Natural History of the Piranha

The name 'piranha' conjures up an image of a steaming Amazon River in which a hapless animal straying into it suddenly disappears underwater in a broiling mass of blood. A few minutes later, all is calm – the only trace remaining of the animal is a skeleton clinically stripped of its flesh. This was the type of description given by U.S. President Theodore Roosevelt in his 1914 book *Through the Brazilian Wilderness*. He stated 'They will snap the finger off a hand incautiously trailed in the water...They are the fish that eats men when it can get the chance...Blood in the water excites them to madness...'. However, this spectacle could have been purposely orchestrated for him with a shoal of piranhas held in the river without food for some time. This episode convinced Roosevelt that piranhas are 'the most ferocious fish in the world'.

The fearsome reputation of the piranha is matched today only by that of the great white shark, all the more impressive given the piranha's diminutive size (usually between 15 and 25 cm in length; Figure 1.1) compared to several metres of shark. Unlike the shark, piranhas usually attack prey larger than themselves.

The piranha's aggressive image has proved highly marketable, giving its name to, among other things, a thirst-slaying, fizzy drink with its 'bone-crushing citrus' (Figure 1.2); a U.S. women's ice hockey team, the 'Pittsburgh Piranhas'; and two well-known rock bands.

Piranhas have also inspired Hollywood. In the 1978 cult horror film *Piranha*, a shoal of piranhas, genetically engineered by the military to be oversized, fast-breeding and flesh-crazy, is accidentally released into a river leading to a holiday resort, with predictably gory results. In the 2010, 3-D film *Piranha*, a shoal of piranhas thought to be extinct for over 2 million years is released from an underground lake into a river following an earthquake. They reach a town where a large number of teenagers, swimming in the river, meet with a violent end. In the James Bond film *You Only Live Twice*, supervillain Ernst Stavro Blofeld feeds one of his female agents to his pet piranhas as a punishment for failing to kill Bond.

Much ferocious piranha behaviour has been wildly exaggerated. Of the more than 30 different species, many do not kill other fish, but merely 'graze' off them, taking small pieces of the fin or scales, which will later regenerate. Some are vegetarians. Only a few species are considered fierce and aggressive. Attacks on humans are extremely rare, and deaths are virtually unknown. Amazonian children swim quite safely in streams and rivers inhabited by piranhas, as has the occasional

Nothing but the Tooth. DOI: http://dx.doi.org/10.1016/B978-0-12-397190-6.00001-8

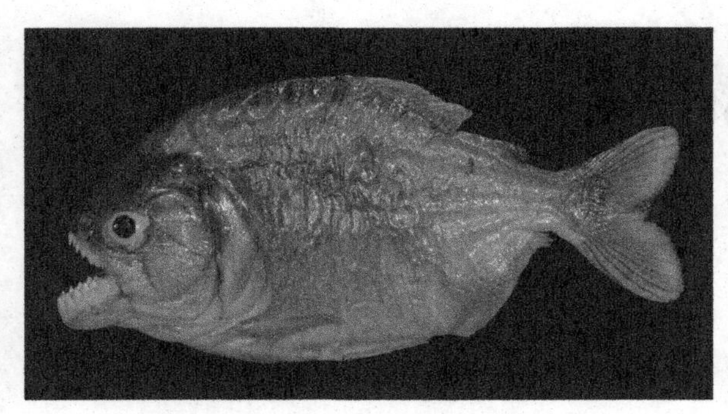

Figure 1.1 A 19-cm long piranha showing the general body features. Note the deep-bodied outline, the forward-jutting lower jaw and the wide gape of the jaws.

Figure 1.2 A can containing a fizzy drink called Piranha with the logo 'bone-crushing citrus'.

intrepid, documentary filmmaker. A recent exception occurred in 2002 in southeast Brazil, where over 50 attacks were recorded (the most serious requiring a toe amputation). This unusual aggression seemed to be a consequence of damming the local river for flood protection, resulting in a large number of understandably irritable piranhas being trapped in the still waters, compared with the flowing white waters of their native environment.

The giant feeding frenzy is also a myth: some piranhas hunt individually, whereas shoals typically contain only 20–30 individuals. Piranhas tend to pair up when swimming in shoals, and if there is an odd number of fish, the unlucky unpaired single one may be eaten by its fellows. Although previously regarded as an attacking strategy, recent research has suggested that shoaling behaviour may also represent a defence against predators (i.e. safety in numbers).

Piranhas are bony fish belonging to the genus *Serrasalmus*, meaning saw salmon and referring not to the teeth but to the notched appearance of the enlarged scales running in the midline across the belly. The origin of the word piranha is uncertain but may be derived from a local Amazonian tribal word meaning either toothed fish or scissors. In support of the latter, piranha jaws are used as cutting or shearing tools, even being employed to cut hair or to sharpen poisonous darts.

Although their main prey is other fish, piranhas are opportunistic and will eat whatever is available. Analysis of their stomach contents reveals insects, crustaceans, molluscs, lizards and rodents. As the rivers flood into the forests, piranhas incorporate seeds and other vegetable matter into their diet. Like carnivorous predators on the African savannah, piranhas seek out and dispatch the weak and injured animals in Amazonian rivers. Their extreme sensitivity to the scent of blood helps them in this pursuit. Birds, such as young herons and egrets, falling from their nests into the river below supplement the piranha diet.

Piranhas themselves have many predators such as larger fish, birds, caiman, freshwater dolphins, turtles and the indigenous people. A new tourist attraction of piranha fishing is being promoted. In the dry season, piranhas may become concentrated in isolated, shrinking, shallow pools that provide particularly rich pickings for fish-eating birds (including herons).

Teeth of the Piranha

The success of piranhas in obtaining food depends not only upon cooperating as a pack, like wolves, but also in their possession of an extremely specialised set of teeth. The unique arrangement of their teeth makes the piranha very effective as a predator. Whereas the majority of fish have large numbers of teeth (in many instances reaching a hundred or more, as in the barracuda; see Figure 4.10), the piranha has an upper row of only six and a lower row of seven on each side, and this number is retained throughout their life. Unlike most fish, whose teeth are conical and spaced out (Figure 1.3), piranha teeth are sharp, triangular, narrow and bladelike with razor-sharp edges and are in contact with each other (Figures 1.4 and 1.5). In addition to the central, prominent, main cusp, there is a small cusp at

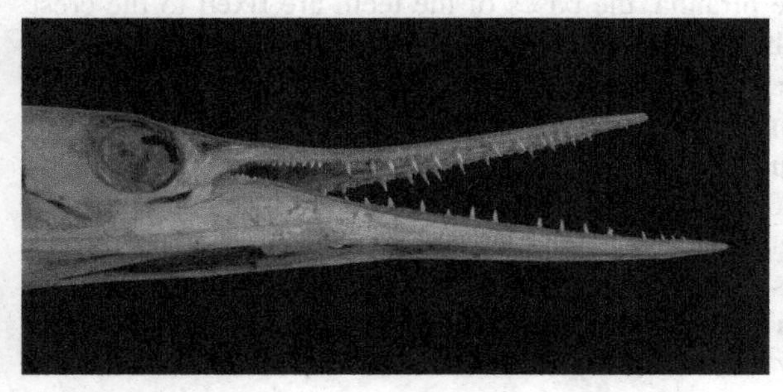

Figure 1.3 The teeth of the short-nosed gar fish. Note the large number of spaced, conical teeth, with gaps along the dentition where replacing teeth will soon erupt.
Source: Courtesy of the Hunterian Museum at the Royal College of Surgeons.

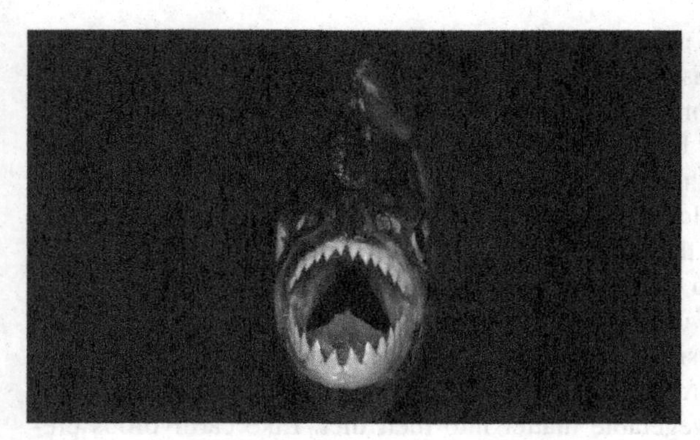

Figure 1.4 View of the mouth of the piranha from the front, showing the razor-sharp pointed teeth.

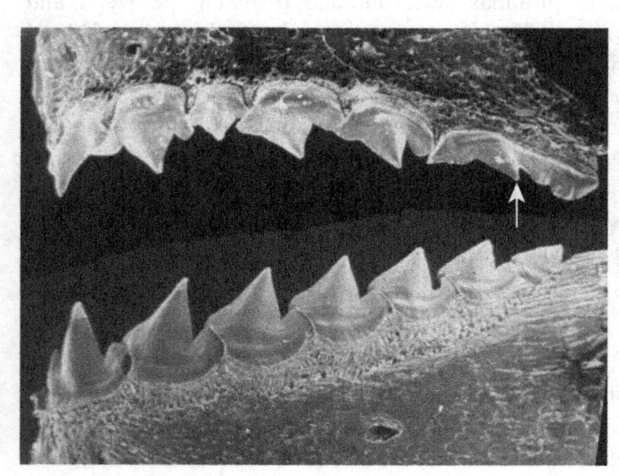

Figure 1.5 The complete upper and lower teeth seen from the left side of the piranha. Note there are six teeth in the upper jaw and seven in the lower. The enlarged upper sixth tooth is arrowed.
Source: From R.P. Shellis and B.K.B. Berkovitz, 1976. Courtesy of the *Journal of Zoology*.

the back of each tooth, which slots neatly into a depression at the front of the tooth behind. This results in the teeth being totally interlocked (Figures 1.5 and 1.6). Instead of a number of separate teeth in each jaw, each piranha jaw could be considered as containing a single, continuous, serrated cutting edge.

In contrast to mammals, whose teeth are secured by roots embedded in bony sockets in the jaws, fish teeth consist only of crowns that are attached to the surface of the bones of the jaw. In the piranha, the bases of the teeth are fixed to the crest of the jaws by a short, inflexible, strong, fibrous ligament. Together with the interlocking of the teeth and the saddle shape of the bone to which the teeth are attached, this provides great firmness and stability (Figure 1.6).

In other species of fish, such as the short-nosed gar, whose teeth are typically conical and spaced apart (Figure 1.3), the upper and lower teeth do not meet, and they are unable to cut up food into smaller, manageable pieces. The teeth function merely to grip the prey, which is usually swallowed whole and head first. A shark can rip chunks of flesh from its prey, but this is achieved more as a result of using twisting movements of its body once the prey has been grasped between its sharp teeth.

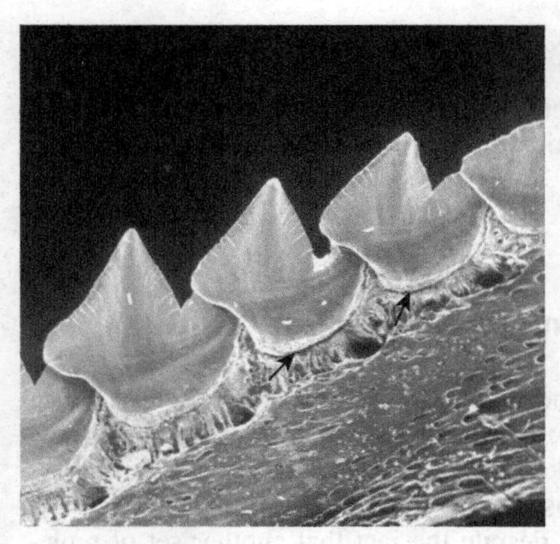

Figure 1.6 High-power view of Figure 1.5 showing the interlocking of the back four teeth in the lower jaw of the piranha. The teeth are attached to the jaw by a thin, dense fibrous ligament (arrows). The teeth lack roots.
Source: From R.P. Shellis and B.K.B. Berkovitz, 1976. Courtesy of the *Journal of Zoology*.

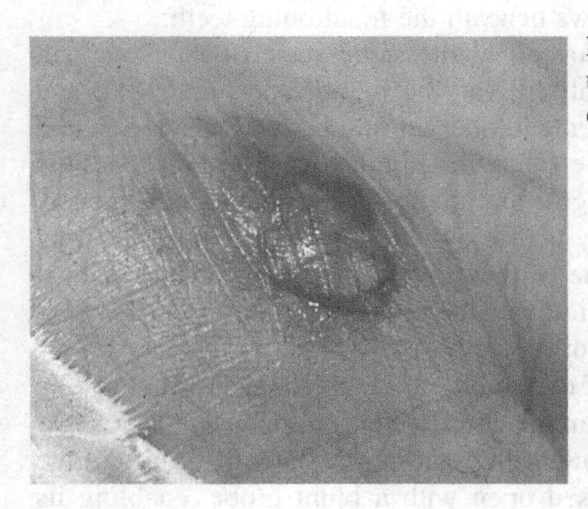

Figure 1.7 The palm of the hand of an unwary aquarist bitten by a piranha. The diameter of the wound was about 1 cm.

The piranha has a characteristic, forward-jutting, powerful lower jaw and is deep-bodied. It has a large gape (the distance between the jaws when the mouth is fully open) relative to its total body length (Figure 1.1); a piranha 10 cm long has a 1 cm gape. Together with its strong jaw muscles, this wide gape helps to generate a very large force when biting. As the mouth is snapped shut, the tips of the lower teeth, which slope increasingly backwards towards the rear of the jaw (Figure 1.5), puncture the prey, and the puncture points are elongated into cuts by the razor-sharp edges of the teeth moving upwards. The back four lower teeth slice against the large, back (sixth) upper tooth, which is greatly expanded in length to form an elongated blade (Figure 1.5), like a pair of scissors. A single bite from a piranha is sufficient to sever a portion of flesh from its prey. This slicing ability, where the teeth shear past each other, is unique among fish.

Figure 1.7 shows the unfortunate result of a piranha bite to a human hand. The individual concerned was transferring a piranha from one tank to another when the fish wriggled free from its net and landed on the floor. Without thinking, the person used their bare hand to push the fish back into the net. The piranha struck instantly, neatly excising a small, circular piece of skin the size of a five-penny piece, with resulting profuse bleeding. When the victim arrived at Accident and Emergency, the doctor, on hearing the cause of the injury, was far from being sympathetic and had difficulty keeping a straight face.

Tooth Replacement in the Piranha

In most fish (as well as in amphibians and reptiles), the teeth are constantly being replaced throughout life (see Chapter 9); as it may take a few weeks for a new tooth to erupt, temporary gaps are seen along the tooth row (Figure 1.3). In observing the skulls of piranhas, however, gaps are never seen. Invariably, the piranha seems to possess a complete set of teeth, despite the fact that another set of replacing teeth is always present within the jaws beneath the functioning teeth.

The replacing teeth in any one jaw are all at the same stage of development, although the precise stage of development will vary between jaws (Figures 1.8 and 1.9). Clearly, there must be tooth replacement, and yet no one had ever reported a situation where teeth were missing in the functional tooth row. The only possible way to explain these findings was to suggest that the teeth are shed simultaneously in complete rows and replaced so rapidly that a jaw without teeth would be rarely encountered in a museum collection. Due to the close interlocking of the teeth, it would be impossible for a tooth in the middle of the row to drop out on its own, as it was mechanically linked to its neighbouring teeth.

The only way to test the hypothesis of extremely rapid replacement was to study live piranhas in an aquarium and to examine their teeth continuously, every few days, over a period of about a year. If caught in a net, a piranha will soon become still and allow its jaws to be gently prised open with a blunt probe, enabling its teeth to be observed and counted (Figure 1.10). The first important observation from this study confirmed the hypothesis that all the teeth in any one row are shed simultaneously. Such a pattern of tooth replacement is unique to the piranha. The second important observation was that the replacing teeth erupt into the mouth

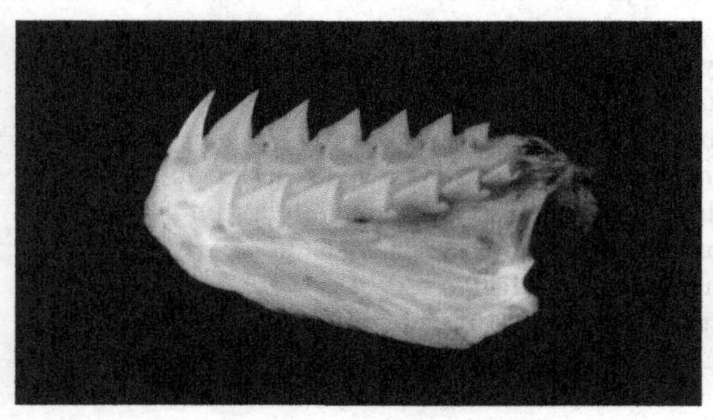

Figure 1.8 An X-ray from one side of a piranha lower jaw, showing the seven functional teeth, beneath which is situated a complete replacing set in which all the teeth are at the same stage of development. *Source*: From R.P. Shellis and B.K.B. Berkovitz, 1976. Courtesy of the *Journal of Zoology*.

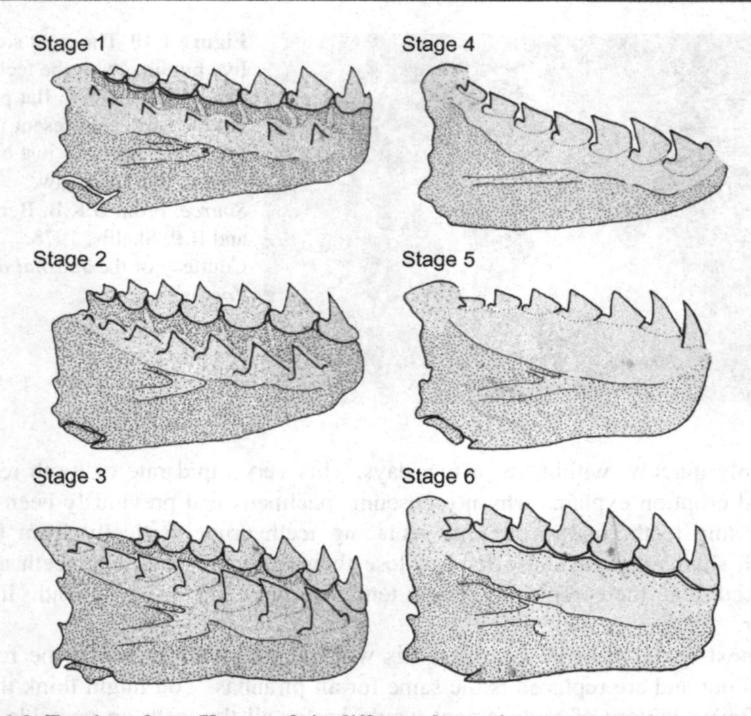

Figure 1.9 Tracings from X-rays of six different lower jaws representing a complete cycle of tooth replacement and illustrating the development, overlapping, eruption and tilting of the teeth necessary to establish the interlocked dentition. Stage 1. With a new set of teeth newly erupted and attached to newly formed bone beneath them, the replacing teeth lying below them are at an early stage of development with just the tips of the teeth mineralised and visible on an X-ray. Note that all the replacing teeth are at the same stage of development.
Stage 2. Mineralisation is more advanced in the developing replacement teeth and has spread downwards from the tips of the cusp to involve the subsidiary cusps at the base of the crown. The two sets of teeth are separated by a reasonable thickness of bone.
Stage 3. The replacing teeth now become tilted in a backward direction so that their open bases face forwards. The sixth and seventh teeth now lie behind the last functional tooth. It is clear that the underlying replacing teeth are erupting upwards as the bone separating them from the functioning teeth above is getting thinner (as it is eaten away, resorbed, by special cells called osteoclasts) and the two sets of teeth are closer together.
Stage 4. The functional teeth have been shed and the bone to which they were attached has been completely resorbed. The replacing teeth have erupted further upwards (as indicated by the increased distance between them and the base of the bony socket in which they were housed). In this stage, the teeth are still tilted backwards.
Stage 5. The final phase in the cycle involves the teeth erupting forwards and rotating into the upright position, thereby ensuring that they interlock correctly.
Stage 6. The teeth then become attached to new bone forming immediately beneath them. A new set of replacing teeth will develop (Stage 1) and the whole cycle is repeated.
Source: From B.K.B. Berkovitz and R.P. Shellis, 1978. Courtesy of the *Journal of Zoology*.

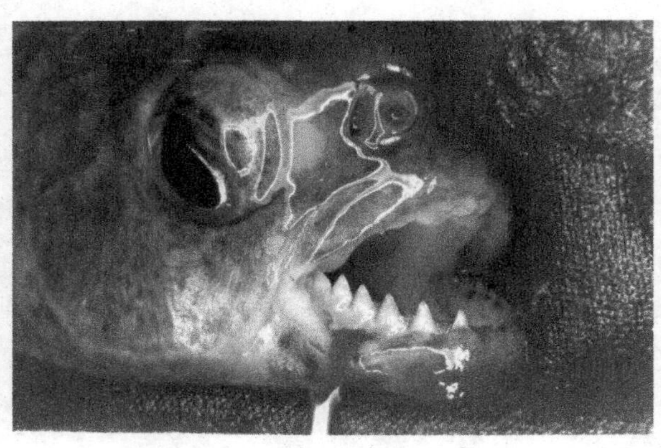

Figure 1.10 The right side of a live piranha, with the teeth exposed by a blunt, flat probe. All the teeth are present in the lower jaw but have just been shed in the upper jaw. *Source*: From B.K.B. Berkovitz and R.P. Shellis, 1978. Courtesy of the *Journal of Zoology*.

remarkably quickly, within just a few days. This very rapid rate of tooth replacement and eruption explains why no museum specimens had previously been found with missing teeth. In the piranha, replacing teeth erupt vertically from below. Although sharks and rays also tend to lose their teeth in rows, their teeth are not interlocked and their replacing teeth tend to move up from behind, like an escalator.

The next question to be considered is whether the order in which the rows of teeth fall out and are replaced is the same for all piranhas. You might think that the most efficient pattern of replacement would be for all the teeth on one side of the jaw to be replaced first, followed later by those on the other side. This would allow the piranha the use of the teeth on one side while those on the other side were being replaced. Indeed, fish with this pattern of replacement were encountered and were seen able to feed. However, many other different patterns of tooth replacement were encountered. One piranha had the upper-right and lower-left tooth rows replaced simultaneously followed later by the lower right and upper left, while in another fish all the teeth fell out at the same time. When meat chunks were dropped into the tank of this particular temporarily toothless fish, it initially attacked the food in its usual manner, oblivious to the fact that it had no teeth. After a few attempts at trying to cut up the food into smaller, more manageable portions, it soon gave up and retreated to a corner of the tank to await the eruption of its new teeth (seemingly with a resigned look on its face!).

The fish whose teeth had all dropped out at the same time might be thought of as disadvantaged (at least temporarily) in the quest to obtain food, as for a short period of time it was completely without teeth. However, the very existence of such a pattern implies that it cannot be a serious disadvantage, as natural selection would have eradicated such a pattern over time. Just as with other carnivorous fish (such as the pike) and many animals, such as snakes, a few days without food would not be a problem for a piranha.

Piranhas grow throughout life, yet the number of their teeth remains the same. To keep pace with jaw growth, the size of the teeth increases very slightly with each new replacement. The number of tooth replacements that occur during the life of a piranha can be estimated using a mathematical formula (known as a Strasbourg plot). For this, it is necessary to measure the length of a tooth (for convenience in measuring teeth from X-rays, the third tooth along the row in the lower jaw) and that of the slightly wider replacing tooth beneath it. These measurements are then related to the length of the jaws in a large series of piranhas, ranging from very young to very old. Such data indicate that a piranha will have had up to 30 sets of teeth in its lifetime. Also with time, the speed of replacement slows down: as the teeth get larger, they last longer. For example, in one medium-sized piranha, the teeth last for about 100 days before being replaced, but 4 years later in the same fish they are replaced after 120 days. Generally, the teeth show little signs of wear at the time they are shed, so that tooth replacement is more related to the increasing size of the jaws than tooth wear.

Is it possible that for some geriatric piranhas, tooth replacement may eventually cease altogether? The probable answer is yes, as one isolated, large, lower jaw has been examined in which the normally sharp teeth were heavily worn down on one side of the jaw, a feature not previously observed. Furthermore, X-rays showed there was no evidence of replacement teeth waiting to erupt beneath these old, worn teeth (Figure 1.11). For this side of the lower jaw at least, tooth replacement had come to an end. On the opposite side of the jaw there were replacing teeth, and the functional teeth had not been worn down.

Another puzzle of the piranha dentition is how the close interlocking of the teeth is achieved. This complex process can be understood by looking at a series of X-rays at different stages of the life cycle of the teeth, from early development to full eruption (Figure 1.9). The replacement teeth in any jaw initially all develop at the same time beneath the functioning teeth, from which they are separated by a bar of bone. They slightly overlap each other and develop with a backward tilt. As they move upwards and erupt, the bone above them is dissolved, eventually resulting in the shedding of functioning teeth. Each new tooth then rotates forwards, commencing from the front, ensuring docking of its small rear cusp within the

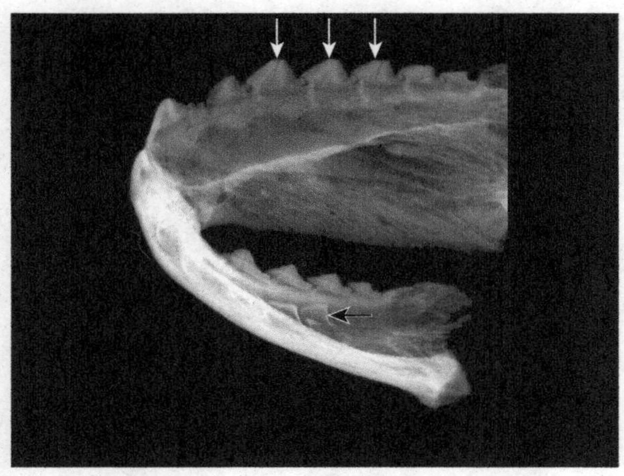

Figure 1.11 X-ray of the lower jaw of a large and old piranha (jaw length just over 4 cm). With no replacing teeth on one side (upper half of image), the functional set has been retained for a much longer period and unusually shows a flattening of the normally sharp cusps (white arrows). The functional teeth on the opposite side (lower part of image) are still sharp and will be replaced, as replacing teeth are present beneath them (black arrow).

scoop-shaped depression of the tooth immediately behind, to establish interlocking. Finally, a layer of new bone is formed below to which this fresh set of teeth becomes attached. A new set of replacing teeth then develops beneath the functioning set and the cycle is repeated.

Summary

Piranha teeth have special features that set the species apart from all other fish. These include (i) interlocking teeth that form a single, functioning, serrated, cutting blade, (ii) the ability of the teeth in opposing jaws to slice past each other and (iii) the presence of very rapid replacement of the whole tooth row. The piranha could be considered to have the second most specialized dentition in the animal kingdom. First prize goes to another, much less glamourous fish, known only to a few specialists and is the subject of Chapter 8.

2 Tusks and Ivory

The hunters gathered together at the entrance to the cave. Their flint-tipped spears lay nearby. It would soon be time to move off and start hunting, for they were hungry. Their quarry was the woolly mammoth, a noble beast that had provided for them for many generations. Hunting the mammoth was dangerous, and many had been killed in the chase. They would need the blessing of the spirits who dwelt in this sacred cave for a successful kill and a safe return. As always, they turned to their shaman, the wise one, who was well versed in the ways of the mammoth. His family alone was favoured with the ability to ensure success in the hunt to follow.

The shaman, his face dyed with red ochre, the colour of blood, was ready to commune with the spirits who controlled the outcome of the hunt. He had also drunk the secret potion made from a mixture of plants. He lit his stone lantern from the fire around the entrance and slowly receded into the depths of the cave. The walls crowded in on him. At some points the channel was so narrow that he had to turn sideways to progress. Suddenly he was in a large recess, and there, in the flickering light of his lantern, he saw on the rock face what he had come for, the ancient images of the animals that shared the land with them. Some were simple outlines in black charcoal, others beautifully depicted in reds, browns and yellows, the topography of the cave surface almost making them spring to life. Every type of creature known to him was there, so well represented as to be instantly recognisable. These noble animals shared the land with them and had equal rights to it. However, some animals had to be killed to provide his people with life's necessities, namely food, clothing, oil for their lanterns and tools. His gaze quickly scrolled over the images of bison, horses, bulls and deer, until it focused on the target of their hunt, the woolly mammoth (Figure 2.1). There it was, near the cave roof, the outline of its body and trunk unmistakable. The shaman knelt down before the image, his thoughts concentrating on the hunt that was soon to follow. In his mind he visualised the hunters, well camouflaged by trees and bushes, slowly following a group of mammoths. They identified an older animal that seemed weaker than the rest and quietly pursued it until they had it surrounded. The bravest and strongest hunter rushed in first and drove his spear into the tough hide, while the others made loud noises to frighten off the rest of the herd. At each opportunity, the hunters drove more and more spears into the wounded mammoth, making sure that it did not escape. Its movements became weaker and weaker until it could no longer stay upright. It collapsed onto the ground. With clubs, they crushed its skull and ended its life. The chase was over and the hunters victorious.

Nothing but the Tooth. DOI: http://dx.doi.org/10.1016/B978-0-12-397190-6.00002-X

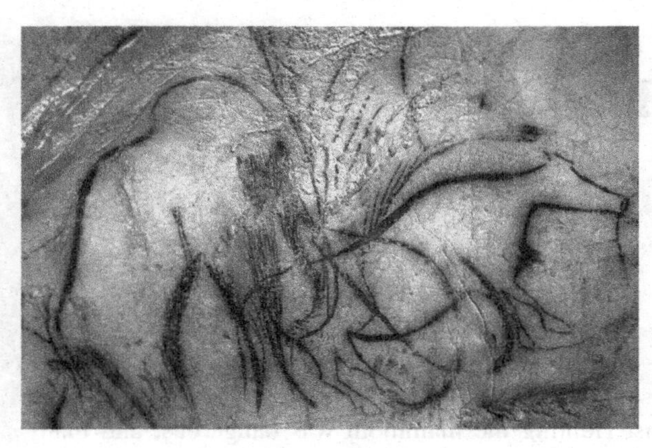

Figure 2.1 Cave painting depicting a mammoth and a cow, from the Chapel of the Mammoths (cave painting), Grotte de Pech Merle, Lot, France/Index/The Bridgeman Art Library.

The shaman came out of his trance, slowly rose to his feet and retraced his steps to the front of the cave. He had been gone for 3 h. The hunters, assured that everything had been done to provide for a successful outcome to their hunt, confidently set off.

A week later the hunters returned, triumphantly carrying meat carved from a woolly mammoth. It was not an old, weakened mammoth that they killed, but a young one that, once they had managed to separate it from its mother, provided an easy kill. Although they left the tusks behind, a few small fragments had broken off and were brought back to present to their shaman. The shaman returned to the depths of the sacred cave to give thanks to the spirits for once more providing them with sustenance.

The shaman kept the small pieces of mammoth tusk close to him. His senses were stirred by the feel of them, and their creamy colour stood out against the drabness of his surroundings. While handling them, he noticed that he could scrape off thin shavings of the tusk with a sharp flint, and that this produced very small changes in shape. One day, while looking at one of the pieces of tusk, he imagined within the small block a figure of the mammoth itself. After months of effort, he had created an exquisite carving of the mammoth, only 4 cm long (Figure 2.2).

The shaman had produced one of the earliest pieces of sculpture in recorded history. It was rediscovered 35,000 years later and created a media sensation. Unwittingly, he also produced an artefact that would endanger the very existence of the type of animal he so much respected and depended upon for food. The material of a mammoth tusk is known to us by another name – 'ivory'.

What Are Tusks?

Not everyone realises that tusks are actually teeth. They are large, curved teeth that project out from the mouth and are supported by roots that are embedded in the jaw. Because of their size, they are among the most highly specialised forms of

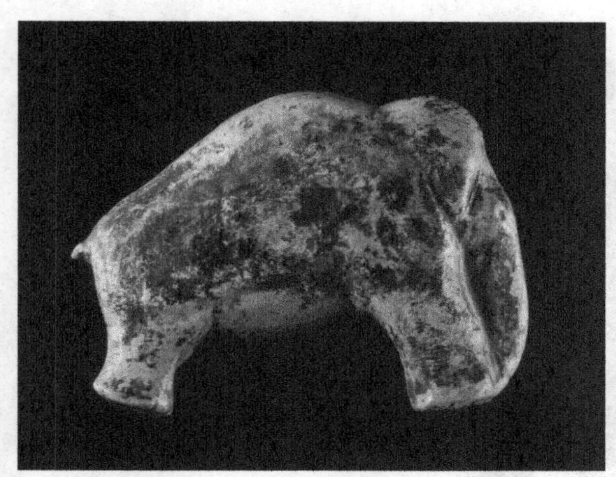

Figure 2.2 35,000-year-old carving showing a woolly mammoth from Germany. Made from mammoth ivory, it is the oldest known ivory carving.
Source: Courtesy of Hilde Jensen, University of Tübingen.

teeth. Tusks reach their great size by growing throughout life. In addition to elephants, tusks are found in hippopotami, pigs, walruses and narwhals (a tusked whale) and are larger in the male of the species.

Unlike other teeth, tusks are unusual in that they generally lack a covering of enamel (except for a small amount at the tip in the newly formed tusk). This means a tusk is composed entirely of dentine, apart from a thin covering of cement, which anchors the tusk to the jawbone. The most recognised tusks belong to the elephant, and the dentine of elephant tusks is usually referred to by the more familiar name of ivory. The same name is also given to the dentine forming the smaller tusks of other mammals as well as the teeth of sperm whales. Unfortunately, man covets and prizes the tusks of elephants, and wars have been fought over obtaining them.

Elephant Tusks

Elephant tusks represent their upper incisors, one on each side. In African elephants, they can reach up to 3 m in length, be up to 20 cm thick, and each weigh up to 90 kg (200 lbs) (Figure 2.3). One-third of the tusk lies hidden and embedded within the socket of bone, with the remaining two-thirds being visible on the outside.

Figure 2.4 shows a diagram of an elephant tusk. The embedded part of a tusk has a soft, central pulp cavity surrounded by hard dentine. The pulp cavity is widest at the base of the tooth but gradually narrows and disappears at the margins of the bony socket as it becomes filled in with dentine, which is formed continuously throughout life. A thin, fibrous joint (periodontal ligament) connects the outer layer of cement to the wall of the bony socket. The exposed part of the tusk is solid and has considerable strength and elasticity.

Figure 2.3 Illustration showing two very large elephant tusks.
Source: Courtesy of the Natural History Museum, London/Science Photo Library.

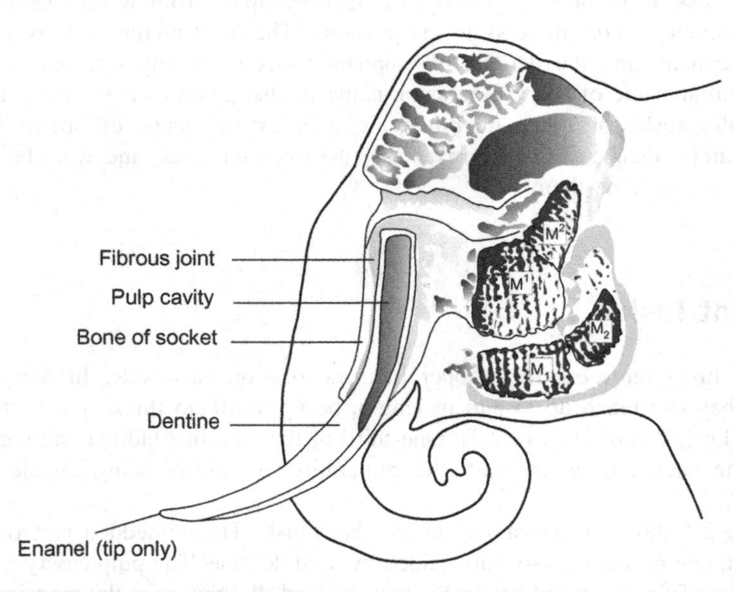

Figure 2.4 Diagram showing the structure of an elephant tusk embedded in the upper jaw. Note that only one pair of grinding molar teeth are ever present on each side of the jaw (labelled M^1 in the upper jaw and M_1 in the lower jaw). When they are worn down, they are replaced from behind by the next pair of molar teeth (labelled M^2 in the upper jaw and M_2 in the lower jaw). When the sixth and last molar teeth have been worn down the elephant would die.

The tusks of African elephants are larger than those of Asian elephants. Tusks can grow up to a rate of 3 mm per week, which, coincidentally, is the same as that for incisors in rodents such as the rat. The difference is that rodents continually wear their teeth away as they use them for gnawing, so that the incisor remains more or less the same size throughout life. Rodents also live only for a year or two, compared with 50 years or more for an elephant, which does not wear its tusks away.

The evolution of tusks has given elephants a remarkable tool to aid in their survival. In both African and Asian species, tusks are used in rooting for food and breaking the bark off trees. Together with the trunk, tusks can help uproot whole trees on which the elephant feeds. They are also used as offensive weapons. The strongest male (bull) elephants with the largest tusks obtain dominance over their rivals in the mating season, hence the name tuskers. Baby elephants have very small milk tusks that drop out after the first year, soon to be replaced by the permanent tusks. Elephants may favour using the tusk on one side.

Elephant Ivory

Because of its size, composition and unusual structure, ivory may be cut to various thicknesses, including very thin layers that allow for delicate inlay work. Most importantly, it can be carved very finely and its surface can be polished, revealing a range of colours from white (freshly cut) to yellow (aged). Together with its sensuous feel, these properties appeal to man's aesthetic nature and, for thousands of years, ivory has been used to produce decoration of all types, from personal ornamentation to works of art, especially religious. In more recent times, ivory provided material for piano keys, cutlery handles and even bagpipe mounts, while its elasticity was highly suited to the manufacture of snooker and billiard balls. Ivory is also an ingredient in some traditional medicines.

Bearing in mind how much ivory in all its forms is in circulation, it is impossible to visualise the vast numbers of elephant tusks needed to supply this demand. Some ivory would have been obtained after the natural death of the elephant, but a large amount, particularly recently, has been obtained following purposeful slaughter. The finest ivory, from African elephants, is more translucent and has a yellower appearance than Asian (Indian) ivory, which is more opaque, whiter and softer.

To understand the properties of ivory, you first need to know a little about its structure. Like other animals, dentine in the elephant tusk is formed by living cells as a composite of organic fibres, which provide the elasticity, and inorganic crystals of calcium phosphate, which give it strength and hardness. Dentine is permeated by millions of minute tubes (dentinal tubules) running from the central pulp outwards towards the surface. Elephant ivory can be distinguished from all other types by the presence of a unique pattern of structural lines. In a cross-section of a tusk, two intercrossing systems of curved lines are seen, referred to as Schreger lines. Superimposed on the lines is a curved, checkerboard arrangement of light and dark squares (Figure 2.5). More detailed microscopic analysis provides an explanation for this appearance (Figure 2.6). The dentine tubules making up the

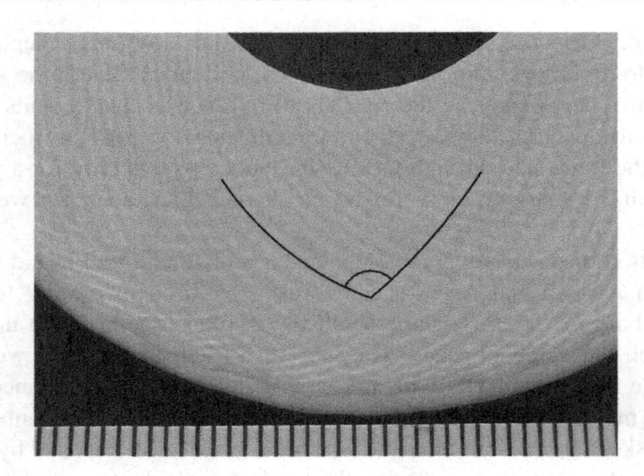

Figure 2.5 Transverse section of elephant ivory showing the characteristic checkerboard pattern. Two lines of Schreger have been superimposed, and the angle of their intersection helps to differentiate the type of ivory. In elephant ivory the angle is greater than 115°, while in mammoth ivory the angle is less than 90°.
Source: Courtesy of Dr S. O'Connor, University of Bradford and the Hunterian Museum at the Royal College of Surgeons.

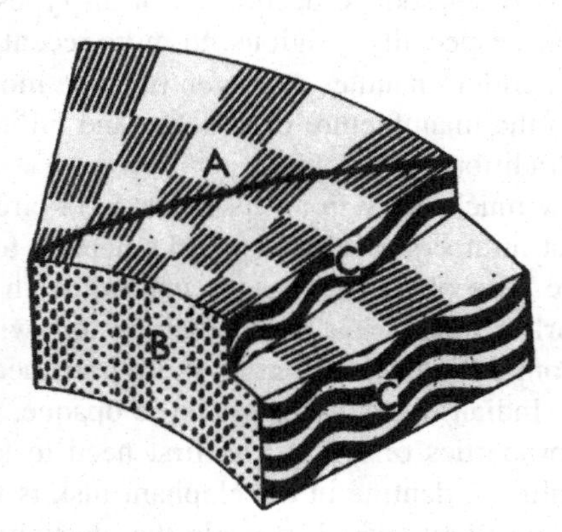

Figure 2.6 Schematic drawing of a block of elephant ivory showing the dentinal tubule pathways. The top surface (A) is a true cross section of the tusk and shows a curve of a Schreger line passing through the dark squares of the checkerboard appearance. (B) indicates the pulpal surface where the dentinal tubules will be seen in cross-section. Part of a column has been removed from the radial surface (C) showing the undulating curvatures of the tubules, with the troughs of the curves in one column being opposite the crests in the adjacent column.
Source: From A.E.W. Miles and J.W. White, 1960. Courtesy of the editors of the *Proceedings of the Royal Society of Medicine*.

ivory are arranged in small groups, within which the undulating curvatures are all in the same phase but are in the reverse phase from those of adjacent columns. In this manner, the troughs of the waves of one column are opposite the crests of the waves in the adjacent column. The ivory will appear dark or light depending on whether the light rays hit the crests or the troughs. The cross-hatching or checkerboard arrangement will not be seen in ivory cut in other planes, where the ivory may have a simpler, striped appearance. Ivory from extant elephants can be distinguished from extinct mammoths according to the angle of intersection of the Schreger lines, that in the elephant being larger.

In addition to its checkerboard appearance, ivory can also be distinguished by viewing it under ultraviolet light. True ivory will fluoresce a bright blue, while synthetic ivory will appear a dull blue. Bone can be readily distinguished from ivory as it lacks the checkerboard appearance and possesses slight imperfections related to the presence of small blood channels.

The poaching of African elephants for their ivory, as well as the recent loss of their natural habitat due to the human population explosion, has led to a great decline in their number. This has particularly affected the big male elephants with the largest tusks, which also provided trophies for rich hunters. With the disappearance of such prize specimens, it seems that residual male elephants with smaller tusks have been more successful in mating with females. This might explain the apparent overall reduction in tusk size in modern African elephant populations. Tusk reduction has been even more extreme among Asian elephants, which have been domesticated. Here, half of the bulls now have no tusks at all, being termed 'makna' or 'mukna'.

To try to prevent the slaughter of elephants, the Convention on International Trade in Endangered Species (CITES) has outlawed the trade in African ivory, except for ivory artefacts known to predate June 1947. All modern ivory artefacts now require strict certification, resulting in some recovery in elephant numbers. This in itself may pose a problem where their habitats are limited and overgrazing may occur.

The ban on illegally traded, modern, African ivory does not apply to prehistoric ivory from related, extinct species, namely mammoths and mastodons. These species were adapted to the cold of the Ice Age. The more common woolly mammoths appeared about half a million years ago and are clearly recognisable in prehistoric cave paintings (Figure 2.1). The rarer American mastodons appeared earlier, about 4 million years ago and, like the woolly mammoth, survived until about 7000–10,000 years ago, so that both species were contemporaneous with modern early man. The tusks of these prehistoric species were longer than those of modern elephants, reaching up to 5 m, and this was further exaggerated by their smaller bodies, an adaptation to the cold. The tusks of mastodons were relatively straight compared with the curved form of mammoths (Figure 2.7). Indeed, mammoth tusks almost contacted each other at their tips.

Huge numbers of tusks of woolly mammoths have entered the ivory market during the last 100 years, particularly from Siberia and Alaska. Global warming and the gradual thawing of the frozen, northern parts of Russia and Canada are likely to expose great numbers of tusks in the foreseeable future. At present, such remains

Figure 2.7 Skeleton of a Columbian mammoth in the George C. Page Museum, Los Angeles.
Source: Photograph by S. Wolfman, 2009. http://en.wikipedia.org/wiki/File:Columbian_mammoth.JPG. Permission to reuse granted under the GNU Free Documentation License http://en.wikipedia.org/wiki/GNU_Free_Documentation_License and the Creative Commons License, http://creativecommons.org/licenses/by-sa/3.0/deed.en

can be claimed by anyone, so that potentially valuable scientific information is lost as tusks are removed by ivory hunters. Although these tusks have been buried in the frozen ground (permafrost) for thousands of years, their ability to be carved remains unchanged. Moreover, this ivory can take on a wider range of colours, which is related to absorption of minerals, particularly iron phosphates, from the surrounding earth. This results in a brownish or a rarer, bluish, surface.

Although the word mammoth has entered everyday language as meaning something of great size, dwarf mammoths the size of baby elephants have also been discovered, one species inhabiting the island of Crete. In such an isolated environment, natural selection favours smaller individuals that can exist with reduced food intake. A similar situation has been seen with the recent discovery of a dwarf human species, *Homo floresiensis* (see Chapter 13).

Tiny particles from meteorites have been found embedded in the tusks of some Alaskan mammoths. Because of their unusual chemical composition, such as their iron/nickel/titanium ratio, these particles are unlikely to derive from any terrestrial source but must come from outer space. As the tusks in which they were found were estimated by radiocarbon dating to be about 32,000 years old, scientists have concluded that a meteorite of unknown dimensions collided with the earth at that time, accounting for the particles embedded in the tusks.

There has been a worldwide adoption of ivory in human civilisation. In the Bible, King Solomon's throne is described as being made of ivory overlaid with gold. Ivory ornamentation adorns grave goods from the tomb of Tutankhamen, dated around BC 1323. The British Museum has a fine collection of carved ivory from the Kingdom of Assyria dating to the BC ninth and seventh centuries.

One of the seven wonders of the ancient world was the 12-m-tall statue of Zeus at Olympia where, in honour of the god, the Olympic Games were held every 4 years, commencing about BC 776. The statue, sculpted by Pheidias, was mounted on a huge base and reached the ceiling of the Temple of Zeus where it was housed. It portrayed Zeus seated on his throne, wearing a robe made of gold. Large amounts of ivory were used to represent areas of the skin of Zeus not covered by the gold robe. A shallow pool was placed in front of the statue to reflect light and to highlight the whiteness of the ivory. Later, the statue was removed to Constantinople but was reputedly destroyed by fire in 462 AD. Remarkably, excavations near the site of the Temple of Zeus have produced evidence of the original workshop of Pheidias, and among the objects recovered was an elephant tusk.

The wealthy, sixth-century Italian Archbishop Maximianus possessed a magnificently carved ivory chair now housed in the Museo Arcivescovile in Ravenna (Figure 2.8). A high point in ivory carving in Europe occurred during the thirteenth

Figure 2.8 Ivory chair of Bishop Maximianus (Italy). (A) Front view. (B) Side and back view. *Source*: (A) is the courtesy of the Lessing Photo Archive and (B) from http://it.wikipedia.org/wiki/File:Cattedra_vescovile_di_Massimiano,_vista_laterale.JPG. Image is in the public domain http://en.wikipedia.org/wiki/Public_domain

Figure 2.9 John Grandisson triptych from the British Museum: this carved ivory was made between 1330 and 1340. It shows the coronation of the Virgin in the centre panel, set above the crucifixion. St Peter and St Stephen are represented on the left wing, St Paul and St Thomas Becket on the right.
Source: http://commons.wikimedia.org/wiki/File:Britishmuseumgrandissontriptych.jpg. Image covered by the Creative Commons License http://creativecommons.org/licenses/by-sa/3.0/deed.en

and fourteenth centuries AD, where it was widely used to portray biblical scenes (Figure 2.9) and to decorate boxes and book covers.

In Japan, carved ivory has always been highly prized, symbolising authority. It was also used for personal items such as seals and netsukes (small, delicately carved ivories used as toggles to hold objects suspended by a cord from a belt).

Other Animals with Tusks

Although strictly speaking only the dentine of the elephant tusk is regarded as true ivory, carvings have been made from the tusks of hippopotami, walruses and wild

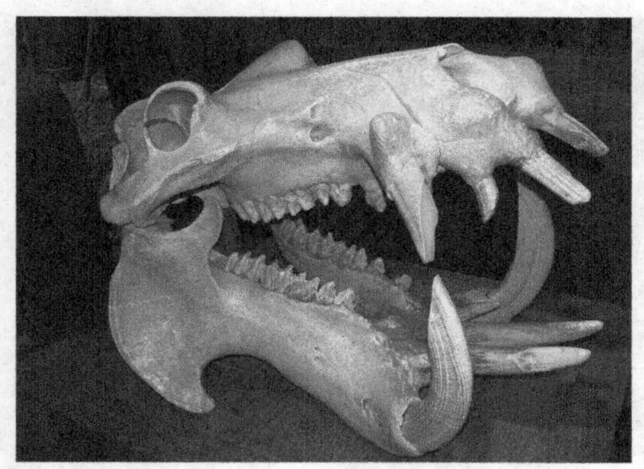

Figure 2.10 Hippopotamus skull showing tusks in both upper and lower jaws.
Source: Photograph taken at Disney's Animal Kingdom by Raul654, 2005. http://commons.wikimedia.org/wiki/File:Hippo_skull_dark.jpg

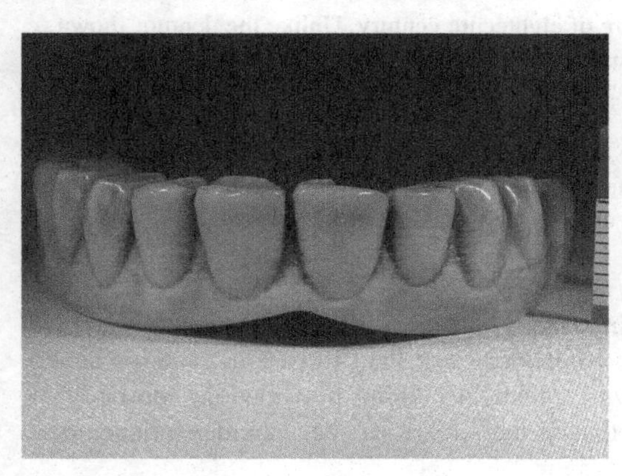

Figure 2.11 Hippopotamus ivory denture made in the eighteenth century. The striped appearance especially prominent on the teeth reflects the highly ordered pathways of the dentine tubule.
Source: Courtesy of Sonia O'Connor, University of Bradford and the Hunterian Museum at the Royal College of Surgeons.

boar. However, their tusks are smaller and cannot be as easily or as finely carved as true ivory. They can be distinguished in structure from true elephant ivory as they lack its characteristic checkerboard appearance and Schreger lines.

Hippopotamus Tusks

The hippopotamus has large tusks in both jaws, representing the upper incisors and the lower canines. The largest and most tusk-like teeth are the lower canines (Figure 2.10), and they are used as offensive weapons. The canines can be up to 1 m long and have a strip of enamel on one surface. Hippopotamus ivory is harder than elephant ivory and is more difficult to carve. It was, however, commonly used in the eighteenth century to make dentures and also to replace individual teeth. The underlying arrangement of the dentinal tubules had the optical effect of producing visible striations (Figure 2.11). In addition to being expensive and with an unrealistic brown colour, the resulting dentures were ill-fitting and unaesthetic and would have become stained and developed an unpleasant smell. In some cases, with skilful workmanship, the thin outer enamel covering on the tusk could be retained as a

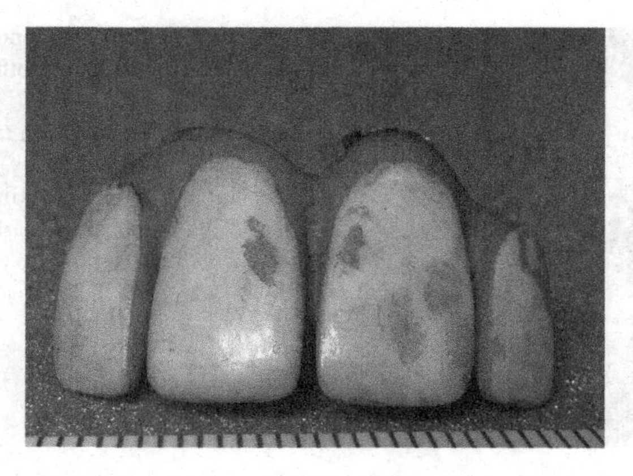

Figure 2.12 Hippopotamus ivory denture of eighteenth century. Unlike the denture shown in Figure 2.11, the enamel has been retained on the surface, giving the front teeth a more natural, white appearance. This rare specimen was identified by Dr S. O'Connor as part of her postdoctoral study, *Cultural Objects Worked in Skeletal Hard Tissues*, which is funded by the Science and Heritage Programme, a joint initiative of the Arts and Humanities Research Council and the Engineering and Physical Sciences Research Council.
Source: Courtesy of the Hunterian Museum at the Royal College of Surgeons.

covering for the front teeth, giving them a more life-like, white appearance (Figure 2.12). It was probably this type of ivory that Paul Revere used in his dental practice to replace teeth and that was crucial in helping him identify the body of his friend Dr Joseph Warren, killed during the American War of Independence (see Chapter 15).

Walrus Tusks

The walrus inhabits Arctic waters. Its tusks are the upper canines and can grow to about 1 m in length, being longer in males (Figure 2.13). Males use their tusks in fighting other males for dominance over breeding. Tusks are also used to make and maintain holes in the ice and for helping the animal haul itself out of the water onto ice floes.

Walrus ivory was the form native to Europe. It was carved by Scandinavians and Inuit peoples. As well as being used for decorative objects, it was also a component of tools and weapons, such as harpoon tips. As with mammoths and mastodons, prehistoric walrus tusks have been collected from frozen ground in Arctic regions and used as a substitute for elephant ivory. The outer (primary) layer of walrus ivory is like elephant ivory, but walrus tusks are less suited for carving. This is because they are much thinner, have a thick covering of cement and a central zone of dentine that has a darker, marbled appearance (Figure 2.14).

Among the most important artefacts carved from walrus ivory are 93 chess pieces discovered on the Isle of Lewis in the Outer Hebrides and known as

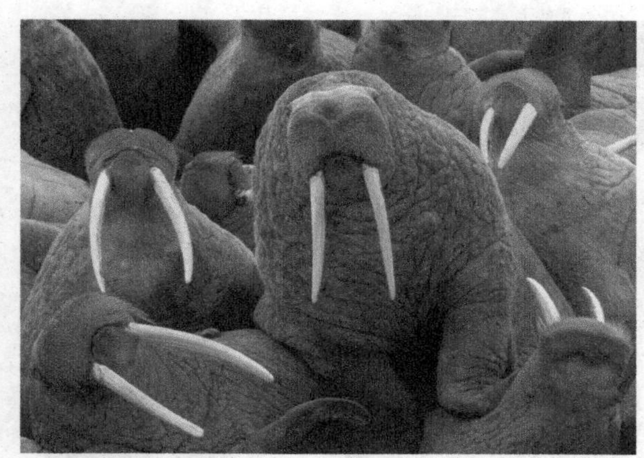

Figure 2.13 The Pacific walrus.
Source: U.S. Fish and Wildlife Service, Pfinge, 2005. http://en.wikipedia.org/wiki/File:Walrus2.jpg. As a work of the U.S. federal government, the image is in the public domain http://en.wikipedia.org/wiki/Public_domain

Figure 2.14 Transverse section walrus tusk showing distribution of tissues. The outer ivory is similar to elephant ivory, but the central ivory is marbled and unsuitable for carving.
Source: Courtesy of the Hunterian Museum at the Royal College of Surgeons.

The Lewis Chessmen (although a few pieces are made from sperm whale teeth) (Figure 2.15). They were unearthed in 1831 and are thought to have been carved in the twelfth century AD in Norway, whose kings ruled that part of Scotland at the time. Eighty-two of the original pieces are exhibited in the British Museum, and the remaining 11 pieces in the Museum of Scotland in Edinburgh.

Boar Tusks

In some wild relatives of domesticated pigs, the upper and lower canines form large tusks. They are curved and project upwards and backwards, the upper tusks being larger than the lowers. Mutual wear creates and maintains sharp cutting edges. The tusks are used for digging up food and as offensive weapons.

In the South Pacific, pigs play a very important role in the culture of certain tribes and are a measure of the wealth and power of a person. Some tribes

Figure 2.15 A selection of The Lewis Chessman from the twelfth century AD. The front row has two pawns at the sides, with the king second left and queen second right. The second row has a bishop in the centre flanked by two rooks. Two mounted knights are at the back in the third row.
Source: ©The Trustees of the British Museum.

deliberately remove the upper tusks in young pigs so that the continuously growing lower tusks have no opposing tooth that they can contact. With time, the lower tusks continue to grow and erupt, forming a spiral tooth that may curve back to embed itself in the lower jaw of the animal (Figure 2.16). The tusks can be removed and worn as armlets to enhance the status of the wearer (Figure 2.17). Tusks were also inserted through the noses of warriors and medicine men. They are so important in the culture that a tusk appears on the national flag of Vanuatu (previously known as the New Hebrides), where it represents prosperity (Figure 2.18). There is even a beer in Vanuatu called Tusker (Figure 2.19).

One member of the pig family has an unusual arrangement of tusks. This is the pig-deer found in Indonesia. In addition to the typical curved canine tusks in the lower jaw, males also possess upper canine tusks that are nowhere near the mouth but erupt out of openings on the snout and curve backwards to end near the eyes (Figures 2.20 and 2.21). These upper tusks can reach lengths of 30 cm. Males may use them when fighting each other. In females, the upper tusks are rudimentary.

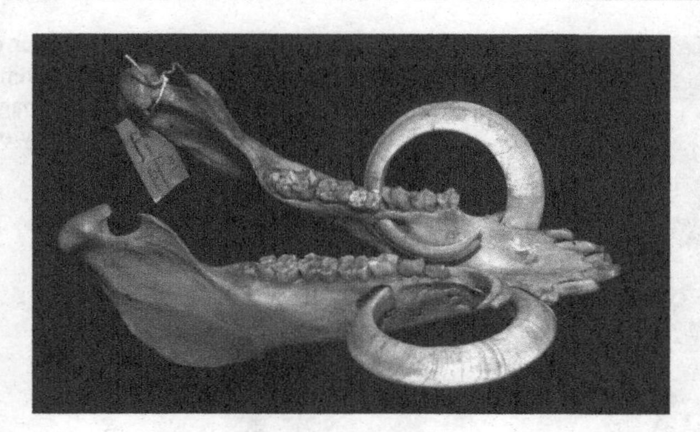

Figure 2.16 Lower jaw of a wild boar in which, following deliberate removal of the upper tusks, the lower tusks have continued to grow unopposed and both have penetrated the jawbone and reemerged. They form an almost complete circle.
Source: Courtesy of the Hunterian Museum at the Royal College of Surgeons.

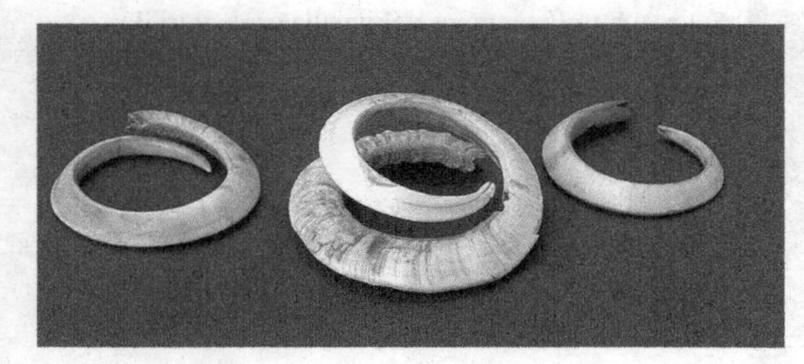

Figure 2.17 Boar tusk amulets formed following the deliberate removal of the opposing upper tusk.
Source: Courtesy of University of Aberdeen.

Figure 2.18 Flag of Vanuatu, showing wild boar tusk near the left border representing prosperity, with leaves of the local fern representing peace. The 39 fronds symbolise the members of the legislature. The Y-shape represents the light of the Gospel going through the Islands.
Source: http://commons.wikimedia.org/wiki/File:Flag_of_Vanuatu.svg. Image is in the public domain http://en.wikipedia.org/wiki/Public_domain.

Figure 2.19 Beer coaster advertising Tusker beer and containing an image of a tusk. *Source*: Courtesy of Dr M. Robins.

Figure 2.20 Male pig deer (Babirusa) showing tusks. The upper tusk grows out of the snout. *Source*: http://commons.wikimedia.org/wiki/File: Hirscheber1a.jpg. This file is licensed under the Creative Commons Attribution-Share Alike 2.0 Germany.

Narwhal Tusks

The most highly specialised of all tusks are seen in a member of the whale family, the narwhal. These whales are found in North Polar regions, around Greenland and in the Canadian Arctic. Not particularly large, with body lengths of around 4–5 m, narwhals have no teeth for chewing and feed on squid, shrimp and other marine animals, which they swallow whole. However, they do have two teeth in the upper jaw. In the male, the left one is a straight tusk which extends from the front of its head for up to 3 m (Figure 2.22). Apart from being straight, its unique distinguishing feature is that the surface has a spiral groove, usually running in a left,

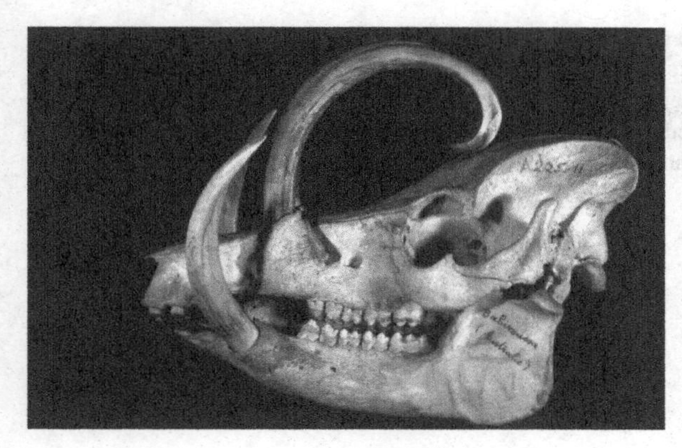

Figure 2.21 Skull of a male pig deer showing tusks.
Source: Courtesy of the Hunterian Museum at the Royal College of Surgeons.

Figure 2.22 Male narwhal displaying tusks.
Source: National Institute of Standards and Technology by author Glenn Williams, 2006. http://en.wikipedia.org/wiki/File:Narwhals_breach.jpg. This work is in the public domain in the United States because it is a work of the United States Federal Government.

anticlockwise direction. Like elephant and walrus tusks, the narwhal tusk has an outer layer of cement covering a core of dentine and lacks a covering of enamel (except sometimes at its extreme tip).

The companion tooth on the right side is small and does not erupt, remaining embedded in the head. Very rarely, however, a male can have two tusks (Figure 2.23). Also, unlike the tusks of other mammals, which are mostly solid, the central part of the narwhal tusk is hollow, containing a long, central, pulp cavity.

The function of the narwhal tusk in males is not completely understood, but it probably gives advantages in breeding either through display or in combat. It has also been suggested that the tusk may be implicated in sound production and reception or as some sort of sensing device. The female narwhal also has incisor teeth that develop in the upper jaw, but both of these are small and rarely erupt.

In the most recent and comprehensive study of the narwhal tusk, it has now been reclassified as a canine tooth (and not an incisor as was generally thought) as it originates in the narwhal's maxillary bone, where canine teeth in mammals originate. Incisors arise in front from the premaxillary bone. The study also reported for the first time the existence of a second pair of small unerupted vestigial teeth located in open tooth sockets in the narwhal's snout alongside the tusks.

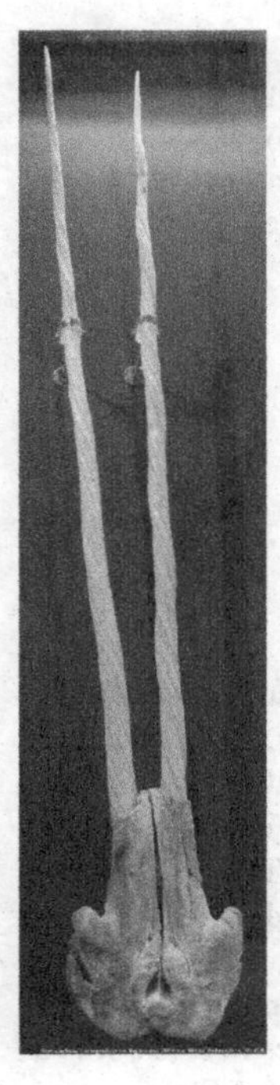

Figure 2.23 Narwhal skull with two tusks.
Source: http://commons.wikimedia.org/wiki/File:Narwalschaedel.jpg. Permission to reuse granted under the GNU Free Documentation License http://en.wikipedia.org/wiki/GNU_Free_Documentation_License and the Creative Commons License, http://creativecommons.org/licenses/by-sa/3.0/deed.en

The shape of a narwhal tusk immediately recalls that of the mythological unicorn, a pure white horse with a similarly shaped spiral tusk growing from the front of its head. According to legend, the shy unicorn inhabited magic forests and was associated with goodness and purity. It could be captured only by an equally pure young virgin.

The mythology of the unicorn was prominent in medieval Western Europe, as illustrated in the six huge tapestries depicting the tale of The Lady and the Unicorn (woven in Brussels between about 1480–1500 and displayed in the Musee de Cluny in Paris). Five of the tapestries represent the five senses: hearing (Figure 2.24), sight, smell, taste and touch. A second set comprising seven tapestries exists showing the hunt and capture of a unicorn and was woven at the same time and place: these are now in the Metropolitan Museum of Art in New York. The tapestries form the subject matter of a historical novel, *The Lady and the Unicorn* by Tracy Chevalier.

You can therefore imagine the wonder and amazement of the public in medieval Europe when 'unicorn' tusks were brought back by seafarers, such as the Vikings, braving the Arctic seas. Streetwise merchants in the know would have been careful to guard the nature of their true origin, wishing to maintain belief in the existence of the unicorn for good commercial reasons. With many different magical powers ascribed to them, the cost of a unicorn tusk was astronomical. They were worth far more than their weight in gold. It was believed that any poison in a drink could be neutralised by contact with a unicorn tusk, so wealthy people had drinking cups

Figure 2.24 Part of a reproduction of the Lady and the Unicorn tapestry: Sense of hearing. The unicorn on the right hand side listens to the Lady and her maidservant playing an organ. *Source*: This is a file from the Wikimedia Commons. //commons.wikimedia.org/wiki/File: Franz%C3%B6sischer_Tapisseur_15.Jahrhundert

Figure 2.25 Bejewelled drinking goblet made from a narwhal tusk. Made by Jan Vermeyen around 1600. *Source*: Courtesy of the Kunsthistorisches Museum, Vienna.

made of narwhal tusks (Figure 2.25). Mary, Queen of Scots (1542–1587) had a piece of a unicorn tusk that she used for that purpose, particularly in view of the threat she posed to Queen Elizabeth I, who eventually had her executed for treason.

The tusk of a unicorn was also believed to ward off disease, so parts of narwhal tusks would be worn as necklaces. Often portions of narwhal tusks were ground up

Figure 2.26 Portrait of Emperor Franz II of Austria painted by Friedrich von Amerling in 1832. The Emperor is wearing the Imperial Crown and is carrying the Royal Sceptre in his right hand.
Source: http://en.wikipedia.org/wiki/File:Friedrich_von_Amerling_003.jpg. This is a file from the Wikimedia Commons and is in the public domain.

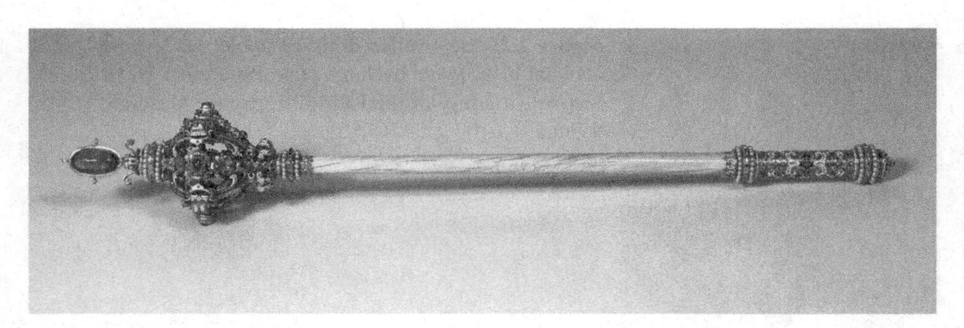

Figure 2.27 The sceptre from the Austrian Crown Jewels. Its handle is composed of part of a narwhal tusk and surmounted by precious stones.
Source: Courtesy of the Kunsthistorisches Museum, Vienna.

and prescribed as a medicine for many diseases. The priceless Imperial Crown Jewels of Austria comprise the crown, orb and sceptre (Figure 2.26). The sceptre, made in about 1615, has a long stem fashioned from the tusk of a narwhal, surmounted by rubies, sapphires and pearls (Figure 2.27). It can be seen in the Imperial Treasury of the Kunsthistorisches Museum in Vienna.

However, by far the most ostentatious display of wealth and power is the Danish Royal Throne (Figure 2.28). Constructed of ivory and numerous narwhal

Figure 2.28 The Danish royal throne constructed of ivory and narwhal tusks, in Rosenborg Castle, Copenhagen. It was made for the coronation of King Christian V in 1671.
Source: Courtesy of the Royal Danish Chronological Collections.

tusks, it is a truly wondrous sight. Housed today in Rosenborg Castle, Copenhagen, it was used for coronations between 1671 and 1840.

It was not until the eighteenth century that the existence of the narwhal and the true source of 'unicorn' horns from narwhals ('sea unicorns') became generally known, after which their price plummeted.

The royal coat of arms of the United Kingdom portrays a shield flanked on the left by the crowned lion representing England and on the right the unicorn, representing Scotland. As a free unicorn was considered dangerous, the heraldic unicorn is chained.

Sperm Whale Teeth

The sperm whale does not have tusks, but it possesses a single set of 40–50 large, conical teeth in the lower jaw, each of which can reach a length of about 20 cm. Although a number of teeth begin developing in the upper jaw, they remain small and fail to erupt. The tooth consists almost entirely of dentine and is considered a form of ivory.

The Inuit peoples inhabiting the Arctic regions of Greenland, Canada and Alaska were the first to develop the art of scrimshaw, which is the carving of scenes on the surface of sperm whale teeth with sharp needles or the point of a knife. Walrus ivory was also used for scrimshaw.

Figure 2.29 Scrimshaw with two sperm whale teeth carved with the portraits of Queen Victoria and Prince Albert surrounding a fob (pocket) watch holder.
Source: Courtesy of Dr J. Papworth.

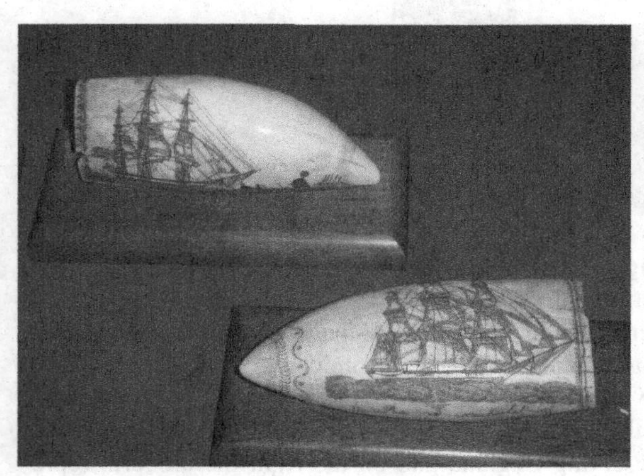

Figure 2.30 Scrimshaw produced by Edward Burdett on two whale teeth.
Source: http://en.wikipedia.org/wiki/File:Nantucket_Rose_scrimshaw_E.Burdett.jpg

Sailors from Europe and America took up the art of scrimshaw to fill idle moments on board whaling ships. Initially, whaling trips were comparatively short (a few months), extending only as far as the coastal waters of the Atlantic, Africa and India, leaving little time for such hobbies. However, once they started hunting in the Pacific, trips could last for 3 years or more. As whaling could take place only during daylight hours, this left ample leisure time in the evenings. The golden age of scrimshaw was the mid-nineteenth century, especially in the American whaling fleet, which had over 700 ships. The sailors cut out scenes on the surface of sperm whale teeth, often just penetrating the cement. Sperm whale ivory cannot be carved too deeply, as the underlying layers tend to be discoloured and the teeth contain large pulp chambers. The outlines of the images were highlighted by staining with dark matter such as tobacco juice or black ink. Any subject might be illustrated, from flags to portraits of family and friends left behind. Figure 2.29 shows two sperm whale teeth engraved with portraits of Queen Victoria and her husband

Prince Albert flanking a pocket watch holder. Whaling scenes with the name of a ship are common subjects for scrimshaw.

As scrimshaw is a very early indigenous American craft, such items are now highly valued. Not many scrimshaws can be attributed to a particular sailor, but when they can, this increases their value. Two names stand out as scrimshaw artists in the American whaling fleet: Frederick Myrick and Edward Burdett. The former produced over 30 examples of scrimshaw from the whaler *Susan*. One of Myrick's scrimshaws sold at auction in America recently for $300,000 (£200,000). Two examples of Edward Burdett's scrimshaw are seen in Figure 2.30. Unfortunately, he drowned at the age of 27 when caught up in a harpoon line that snagged him around his feet and dragged him overboard.

In England, a scrimshaw carved by James Bute, who travelled around the world with Charles Darwin on *HMS Beagle*, and illustrating this famous ship on both faces of the tooth, was recently sold for £40,000 ($65,000).

3 How Two Young Dentists Changed the History of Surgery: Horace Wells (1815–1848) and William Thomas Green Morton (1819–1868)

Dentistry is a relatively specialised, essential, but rarely life-saving service within the medical profession. However, there are a few dentists who have achieved recognition far beyond dentistry. Two dentists to whom this applies are still remembered today for their contribution to general surgery. To appreciate their enormous contribution, it is necessary to remember how operations were carried out up to the 1840s. Successful surgery then was limited to just a few minor operations. A surgeon attempting major procedures knew the very real risk to the patient's life. Firstly, the causes and significance of infection were unknown to the medical profession at that time, and therefore they were unable to prevent them. Secondly, there was no method of pain control (anaesthesia), so that the extreme pain accompanying the surgeon's knife resulted in severe shock to the system and the serious consequences involved.

The necessity for speed in surgery, due to the inability to control bleeding, limited its quality. In order to amputate a limb, as was frequently necessary for war casualties, the patient would be tied down and/or physically restrained by two or three assistants and given a wooden block to bite on. A stiff drink was known to help somewhat at operations, although the large doses required to have any effect were likely to cause vomiting instead of a drunken stupor. A less scientific method occasionally used consisted of a knockout punch to the jaw (as long as any resulting fracture caused less hurt than the operation). In similar fashion, the patient's head was occasionally protected with a helmet, after which a strong blow was struck on it with a wooden hammer!

With the patient inevitably awake and terrified, the surgeon would begin the amputation. The skin and muscle would be cut to expose the bone, which would then be sawn off (generally within about a minute): the wound might then have been cauterised with a red-hot poker or had boiling oil poured into it in the hope of stemming post-operative bleeding. The patient would be fortunate to faint during the procedure. The only certainty along with the excruciating pain was a high death rate. Surgeons of yesteryear must have had very strong constitutions, especially as they did not have the same standing as physicians. It is against this background

Nothing but the Tooth. DOI: http://dx.doi.org/10.1016/B978-0-12-397190-6.00003-1

that two young American dentists, Horace Wells and William Thomas Green Morton, enter the story.

Horace Wells

Horace Wells (Figure 3.1) was born in January 1815 in Hartford, Vermont. He lived in a farming community, receiving a good education at local schools. He moved to Boston at age 19 and chose a career in dentistry. Although some dentists were educated at that time, occasionally even obtaining a medical degree, the majority had no formal qualifications. The reputation of dentists was low, especially compared with doctors. Dentistry was carried out mainly by quacks and charlatans so that, wherever possible, people avoided treatment. Learning the practical skills meant becoming apprenticed to a dentist, observing and assisting. However, improvements for better training were under way with the opening of the first dental college in Baltimore in 1840.

Following a 2-year apprenticeship, Wells opened his own dental practice in Hartford, Connecticut, in 1836: he was 21 years old. At that time, dentistry consisted mainly of extracting teeth, plugging large cavities and providing false teeth. Unlike many in his profession at the time, Wells was an intelligent, enquiring, conscientious and innovative practitioner. In 1838, he published a small pamphlet entitled 'An essay on teeth, comprising a brief description of their formation, diseases and proper treatment (Figure 3.2). In this 70-page essay, the young Wells considers some of the important clinical aspects of dentistry. He reveals himself as mature, well-read and literate, with a considerable mastery of his subject. He displays a sharp mind and evidence of a critical, scientific approach. He scolds the great John Hunter (see Chapter 12) for suggesting the teeth do not have a circulation (viable dental pulp), commenting that 'Mr Hunter was not a practical Dentist: but a very able Surgeon, and writer: had he attended to the practical part of Dental Surgery, it is not improbable that his opinion on this subject would have met with a material change'. In discussing the causes of tooth decay, he makes the following perceptive

Figure 3.1 Portrait of Dr Horace Wells.
Source: http://commons.wikimedia.org/wiki/File:Wells_Horace.jpg. This media file is in the public domain in the United States. This applies to U.S. works where the copyright has expired, often because its first publication occurred prior to 1 January 1923.

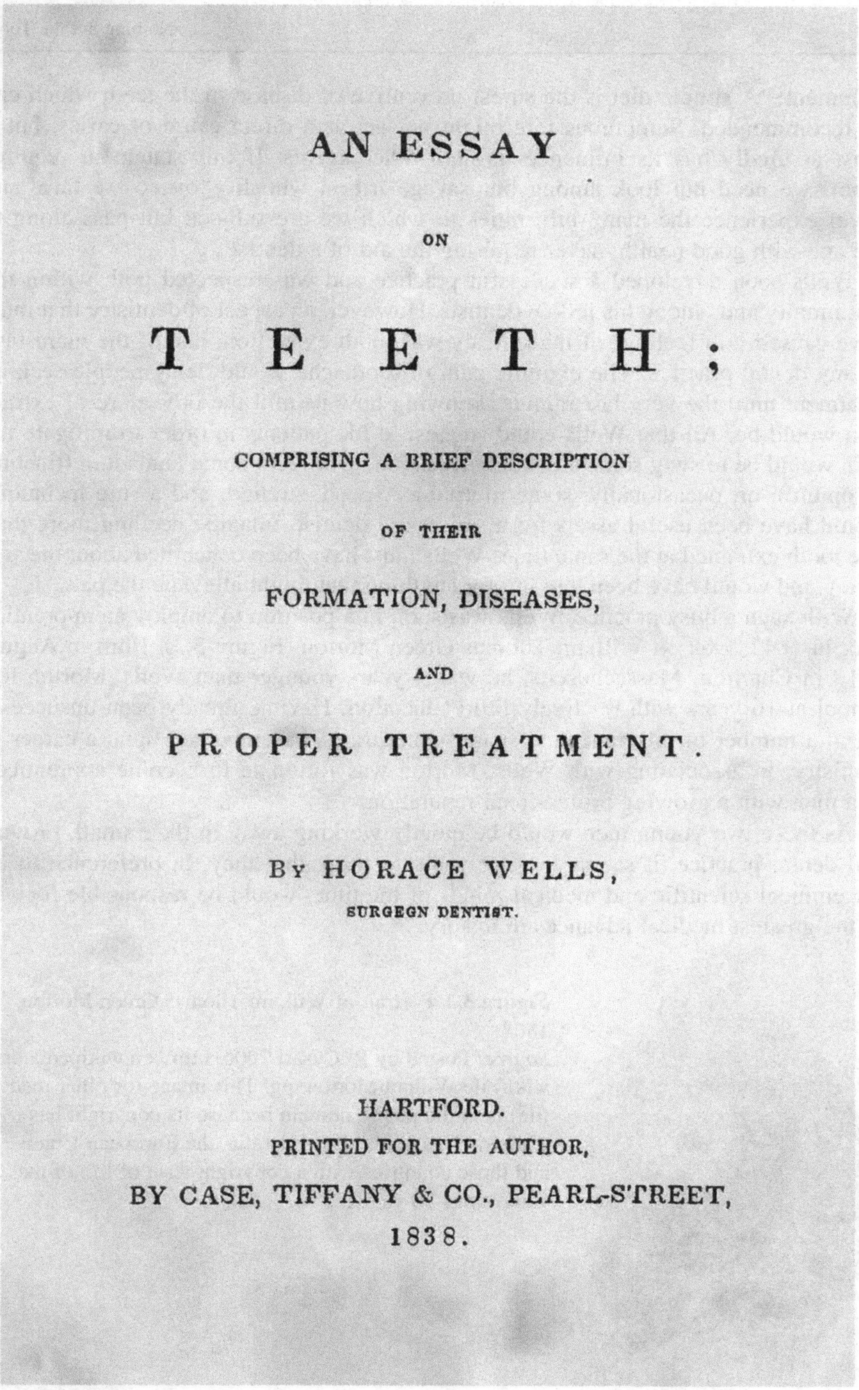

AN ESSAY

ON

TEETH;

COMPRISING A BRIEF DESCRIPTION

OF THEIR

FORMATION, DISEASES,

AND

PROPER TREATMENT.

BY HORACE WELLS,

SURGEON DENTIST.

HARTFORD.

PRINTED FOR THE AUTHOR,

BY CASE, TIFFANY & CO., PEARL-STREET,

1838.

Figure 3.2 Front piece of the 1838 pamphlet produced by Horace Wells.
Source: Courtesy of the U.S. Library of Congress.

statement: 'A simple diet is the surest preventive of disease in the teeth which can be recommended. Sumptuous fare might not act as a direct cause of caries, but it most assuredly has its influence through other agents. If this statement requires proof, we need but look among our savage tribes, who live on coarse fare, and never experience the many infirmities to which we are subject; but pass along to old age with good health, never requiring the aid of a dentist'.

Wells soon developed a successful practice and was respected both within the community and among his fellow dentists. However, an aspect of dentistry that must have caused him feelings of inadequacy was tooth extraction, one of the mainstays of any dental practice. The extreme pain of toothache would delay people seeking treatment until the very last minute, knowing how painful the procedure of extraction would be. All that Wells could suggest to his patients in order to mitigate the pain would be to swig some alcohol beforehand, give them some laudanum (tincture of opium), or, occasionally, some morphine. Speed, strength and a sure technique would have been useful assets for a successful dentist. Imagine needing more than one tooth extracted at the same time! Wells must have been concerned about the suffering and would have been looking for anything that might alleviate the pain.

With such a busy practice, Wells was soon in a position to employ an apprentice and, in 1842, took on William Thomas Green Morton (Figure 3.3). Born in August 1819 in Charlton, Massachusetts, he was 4 years younger than Wells. Morton left school at 16 years with relatively little education. Having already been unsuccessful in a number of different (and shady) enterprises, he embarked upon a career in dentistry. In associating with Wells, Morton was fortunate to become apprenticed to a man with a growing professional reputation.

As these two young men would be quietly working away in their small, provincial dental practice, it seems inconceivable to think that they, in preference to all the eminent scientific and medical minds of the time, would be responsible for one of the greatest medical advances in history.

Figure 3.3 Portrait of William Thomas Green Morton, 1868.
Source: Posted by R. Cusari 2006. http://en.wikipedia.org/wiki/File:WilliamMorton.jpg. This image (or other media file) is in the public domain because its copyright has expired. This applies to Australia, the European Union and those countries with a copyright term of life of the author plus 70 years.

Figure 3.4 Portrait of Dr Charles Thomas Jackson. *Source*: http://commons.wikimedia.org/wiki/File:Jackson_Charles_Thomas.jpg. This media file is in the public domain in the United States. This applies to U.S. works where the copyright has expired, often because its first publication occurred prior to 1 January 1923.

By 1843, the enterprising Wells had developed a new technique whereby he could successfully solder false teeth onto a denture without any subsequent corrosion. As was common at the time, he sought an endorsement of his discovery from a prominent scientist and chose Dr Charles Thomas Jackson (Figure 3.4). Jackson agreed to allow his name to be used in endorsing this product. In the latter part of 1843, Wells and Morton formed a partnership to use this new product, and both were confident that it could make their fortunes. To this purpose, they opened a new practice in Boston.

Unfortunately, the partnership did not immediately bring in the economic rewards expected and was dissolved within a few weeks. There may have been personality clashes; but whatever the cause, the eventual result was that Wells returned to Hartford while Morton stayed on in Boston.

Wells announced his return in the Hartford Courant (Figure 3.5) and there re-established his dental practice. A form of entertainment popular at the time consisted of shows featuring 'laughing gas'. Laughing gas was the common name for nitrous oxide, which had first been identified by the English chemist Joseph Priestly in 1772. In 1800, another Englishman, Sir Humphry Davy, described numerous detailed experiments in which he himself inhaled the gas in varying quantities and for varying amounts of time. He also described the reactions of friends who volunteered to inhale the gas, including the poet Samuel Taylor Coleridge and the potter Josiah Wedgewood. Davy discovered that inhaling the gas aroused him to an excitable state, sometimes leading to uncontrollable laughter, but the effects quickly wore off. For this reason, Davy called nitrous oxide laughing gas. He also wrote with great prescience that, as the gas 'appears capable of destroying physical pain, it may probably be used with advantage during surgical operations in which no great effusion of blood takes place' (Figures 3.6 and 3.7). However, these perceptive comments were not followed up for another 50 years.

At laughing gas entertainments, an entrance fee would be charged and the showman would give an introductory lecture on the properties of the gas. This would then be followed by practical demonstrations when volunteers from the audience would come up on stage and inhale through the mouth from bags containing the gas, while their nostrils were pinched together. When the volunteers had inhaled

t, of fine quality
nue to pay partic-
nd feels confident
s as are not often
I O. PITKIN.
3d&w9

been appointed
district of Hart-
ge Barnard, late
ereby give notice
ppointment at the
in said Hartford
t, and the second
A. M., on each of
d for creditors to
the subscribers
Commissioners.
3d

beautiful *Gold Bosom Pins* and imitation do., of the most beautiful patterns; pure Silver Spoons, all shapes; London and French Cloths; cases and bales of Prints, Sheetings, Shirtings, Ticks, Napkins, Linen Hdkfs., at prices which none *shall undersell.* For the best quality of goods and the most beautiful patterns at the lowest prices, call at the
sept 15 BAZAAR, 259 Main street.

DENTIST

H. WELLS, DENTIST, has resumed his professional business at No. 14 Asylum street, a few doors from Main st. sept 16 d

WANTED—A young lady qualified to work at the Millinery business, to go south. Apply at 235 Main street, 10d sept 10

Hoteling J J
Holbrook, Miss Ma
Holcomb. Burt
House, Wm W
Hollister, Miss Mar
Huntley, Caleb
Hardy, Maria
Hurlbut, Mrs Sinna
Hamilton, Benedict
Humphrey, Miss Ju
Hayden, Isaac S
Harney, John
Higgins, Stephen
Hathway, Amos
Heath, Wm S
Howell, Miss Sarah
Humphreys David
Hills, Henry
Hughes F D
sept 16

Figure 3.5 Advert in *Hartford Courant* advising of Wells's return and reestablishing his dental practice. It states 'H. Wells, dentist, has resumed his professional business at no 14 Asylum Street, a few doors from Main St. September 16'.
Source: Courtesy of U.S. Library of Congress.

enough gas, it would give them a feeling of exhilaration and intoxication, leading them to carry out amusing, involuntary antics, such as running around, fighting, bumping into objects and laughing. There were no unpleasant aftereffects as the gas rapidly wore off. The volunteers would have no recollection of their embarrassing on-stage behaviour, much to the amusement of their friends and onlookers.

On 10 December 1844, Wells attended one of these popular shows when the gas was administered by 'Professor' Gardner Quincy Colton, who himself had studied medicine. Wells was an eager volunteer to inhale the gas. Amidst all the escapades carried out on stage, he noticed that one of the participants, who afterwards sat down beside him, had gashed his knee badly while under the influence of the nitrous oxide and yet had no recollection of any pain. Wells immediately had the inspired thought that perhaps teeth could be extracted without pain using the laughing gas.

That same evening he discussed the idea with his colleague and friend John Riggs, who ran a dental practice nearby. They concluded that such a procedure would necessitate taking a patient beyond the light, excitatory stage seen at shows, to a deeper level leading to loss of consciousness. They did not have any knowledge of either the chemical properties or the biological properties of nitrous oxide, but they were aware that there was a risk of danger and even death. Wells had a troublesome upper wisdom tooth (third molar) that needed extracting, and so he bravely (or recklessly!) volunteered to be the first patient.

The following morning, 11 December, Wells went to his surgery to meet Riggs, whose support was noteworthy, because if anything untoward happened, Riggs was likely to receive some very unfavourable publicity. Colton was on hand to provide a bag of nitrous oxide which Wells held while inhaling from it. Wells soon drifted

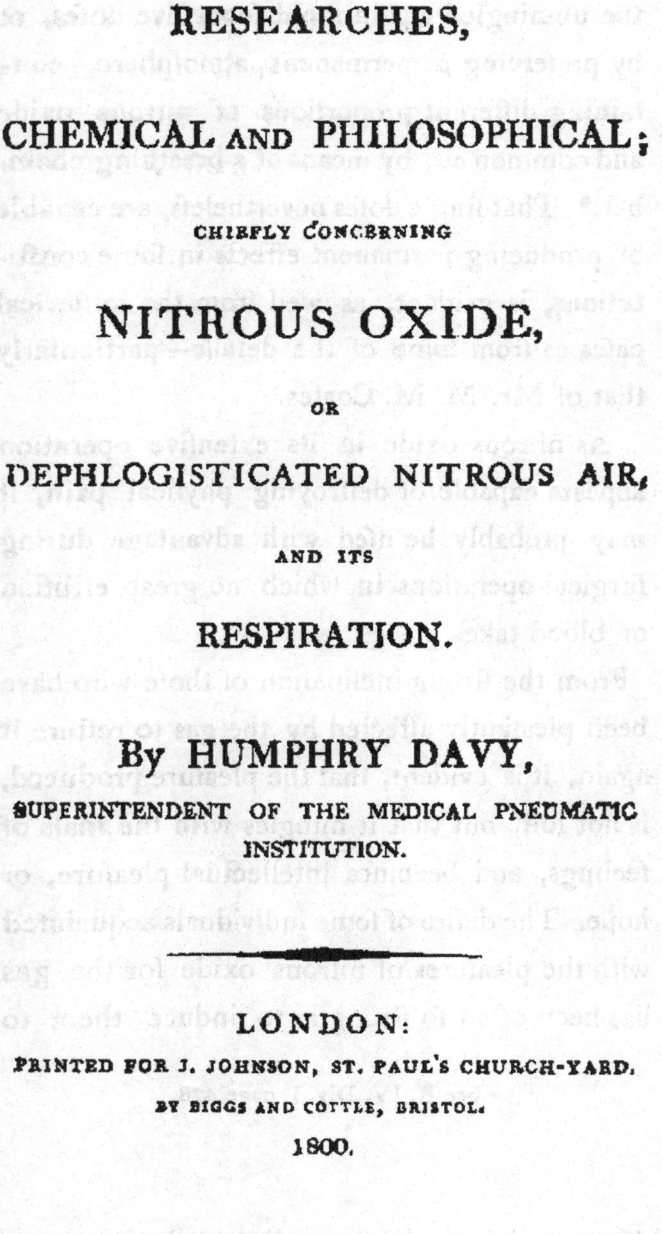

RESEARCHES,

CHEMICAL AND PHILOSOPHICAL;

CHIEFLY CONCERNING

NITROUS OXIDE,

OR

DEPHLOGISTICATED NITROUS AIR,

AND ITS

RESPIRATION.

By HUMPHRY DAVY,

SUPERINTENDENT OF THE MEDICAL PNEUMATIC INSTITUTION.

LONDON:

PRINTED FOR J. JOHNSON, ST. PAUL'S CHURCH-YARD.

BY BIGGS AND COTTLE, BRISTOL.

1800.

Figure 3.6 Front piece to Humphry Davy's book on nitrous oxide published in 1800.

the unmingled gas in rapid successive doses, or by preserving a permanent atmosphere, containing different proportions of nitrous oxide and common air, by means of a breathing chamber.* That single doses nevertheless, are capable of producing permanent effects in some constitutions, is evident, as well from the hysterical cases as from some of the details—particularly that of Mr. M. M. Coates.

As nitrous oxide in its extensive operation appears capable of destroying physical pain, it may probably be used with advantage during surgical operations in which no great effusion of blood takes place.

From the strong inclination of those who have been pleasantly affected by the gas to respire it again, it is evident, that the pleasure produced, is not lost, but that it mingles with the mass of feelings, and becomes intellectual pleasure, or hope. The desire of some individuals acquainted with the pleasures of nitrous oxide for the gas has been often so strong as to induce them to

* See R. IV. Div. I. page 478.

Figure 3.7 Middle paragraph on page 556 from Sir Humphry Davy's book seen in Figure 5. Stating 'As nitrous oxide in its extensive operation appears capable of destroying physical pain, it may probably be used with advantage during surgical operations in which no great effusion of blood takes place.

into a state of unconsciousness, at which time Riggs stepped forward and swiftly extracted the tooth. Recovering rapidly from the anaesthesia, Wells declared he had experienced absolutely no pain. That simple experiment is regarded as one of the great moments in medicine: the first, successful, pain-free operation.

Over the next few days, Wells used nitrous oxide to extract teeth without pain in about 15 patients, generally, but not always, meeting with success. Colton taught him how to prepare and administer the nitrous oxide gas. Hearing about this procedure, local doctors also used the gas for one or two minor surgical operations, with Wells administering the anaesthetic. Wells told others, including Morton and Jackson, of his new discovery.

As Wells believed that an important sign of the anaesthetic effect was its excitatory properties in low concentrations, there is some evidence that he did try using ether, which also had an excitatory phase. In discussing the pros and cons with a medical colleague, Dr Erastus Edgerton Marcy, he was advised that ether was more dangerous and irritating than nitrous oxide, and desisted from its further use.

Wells's discovery was totally revolutionary as he was offering a pain-free state, not by a solid or liquid compound, but by inhalation of a gas. Up to that time, there was a general feeling among surgeons that pain was an inevitable consequence of surgery and could not be dispensed with. In an effort to share his breakthrough with a wider medical audience, in January 1845, Wells went with Morton to the Harvard Medical School in Boston hoping to persuade general surgeons to use nitrous oxide as an anaesthetic.

On 31 January 1845, the Professor of Surgery at Harvard, Dr John Collins Warren, invited Wells to demonstrate his painless extraction method before an audience of medical students at the Massachusetts General Hospital. A student volunteer was placed in a chair and Wells proceeded to anaesthetise him. Unfortunately, when Wells tried to extract a tooth, the volunteer became agitated and reacted as if the procedure was painful, with the result that the demonstration was brought to an abrupt and unsuccessful end. Wells was treated with derision by the students and, with cries of 'Humbug!' resounding in his ears, he fled the scene, bitterly disappointed.

Whether the failure of the demonstration was due to Wells being too anxious and attempting to extract the tooth before the patient was fully anaesthetised, or whether the gas was not pure enough, or whether the patient carried out some involuntary reflex movement and did not experience any pain, is not known. One witness subsequently stated that the tooth was successfully extracted and that there was no manifestation of any pain.

Sadly, nobody offered Wells a second trial. Following his humiliation before the medical profession, which he found so hard to bear, perhaps due to his modest and retiring personality, his health deteriorated. He closed his dental practice and within 3 years was dead. The immediate championing of nitrous oxide anaesthesia expired with him.

At no time did Wells seek to benefit financially from his discovery. He wanted it to be freely available to everyone. The fundamental scientific principle that he discovered first appeared in print in the *Boston Medical and Surgical Journal* on

18 June 1845, in an article written by Dr P. W. Ellsworth. He stated: 'The nitrous oxide gas has been used in quite a number of cases by our dentists, during the extraction of teeth, and has been found, by its excitement, perfectly to destroy pain. The patients appear very merry during the operation, and no unpleasant effects follow'.

William Thomas Green Morton

Returning to the second main character in the story, William Morton (Figure 3.3) stayed on in Boston following the dissolution of his partnership with Wells at the end of 1843. Through dint of hard work, he gradually built up a successful dental practice and soon had students of his own. Initially, he remained in contact with Wells, who continued to provide a source of professional help.

To obtain more status, the ambitious Morton enroled as a student at Harvard Medical School in 1844. At that time, obtaining a medical qualification required satisfactory attendance at lectures and practical classes over a 3-year period and passing various written and oral examinations. Although he attended some classes, Morton never completed the course, mainly due to the necessity of running his dental practice at the same time. During these medical studies, he had a tutor who was none other than Dr Charles Thomas Jackson, the same person who had earlier helped Wells analyse the composition of his solder and endorse the product. Morton and his wife initially boarded with Jackson.

Dr Jackson (Figure 3.4), who was also to play a prominent role in subsequent events, was an internationally recognised scientist. Born in Plymouth, Massachusetts, in 1805, he received his medical degree from Harvard in 1829 and extended his medical studies by sailing to Europe. On the return voyage home, Jackson, always willing to discuss scientific ideas, conversed with a fellow passenger about aspects of electromagnetism. They talked about its potential use in communication. The passenger was Samuel Morse.

In 1834, Jackson constructed a model for telegraphic communication, but not appreciating its full potential, he did not pursue it further. Morse was fully committed to the project and developed a successful communications system whereby a message could be transmitted along a wire. By 1837, he had applied for a patent. The eventual result was the telegraph system, along whose wires signals were sent using Morse code. Jackson felt aggrieved and considered that his ideas had been plagiarised. He spent many years challenging Morse's patent and trying to get credit for the discovery, without success. This type of episode was to be repeated later when he fought for credit in the discovery of anaesthesia.

After his return from Europe, Jackson found less to interest him in medicine and more in the fields of chemistry and geology. He gave up medical practise and instead became a state geologist, achieving considerable eminence.

Sharing lodgings with Jackson allowed Morton lots of opportunities for discussion and advice. At some point, Morton asked Jackson whether he could recommend anything that might help reduce dental pain. Jackson suggested the local application of ether in the mouth. When Morton applied it to a patient, it did seem to provide some relief, although he did not follow-up on its use.

The relationship between the well-educated and well-connected Jackson and the unschooled Morton could not have been smooth for, after a short time, the Mortons moved out and found other accommodation. Occupied with his growing dental practice and having to deal with the ever-present problem of pain during tooth extraction, sometime in 1846 Morton had an idea for a new method of pain control. It might have been the result of one of his apprentices recalling experiences of ether 'frolics', where students sniffed the fumes of ether, giving them a 'high'. It could have been triggered by remembering Wells's initial success with nitrous oxide and that Wells had considered ether but had abandoned it due to its possible toxicity. It might also have been the result of recalling his own previous topical use of ether to reduce toothache following advice from Jackson. Whatever the reason, Morton seems to have conceived the idea that inhaling the fumes of ether might achieve an anaesthetic state whereby teeth could be extracted without pain.

Ether had been available to the medical profession for about 200 years. Inhaling it was known to cause an excitatory state, but it was also given in the form of a medicine to treat stomach complaints. Inhaling was recommended only with caution, and no one had ever considered using it in sufficient concentration to cause loss of consciousness, which is what Morton was now contemplating.

Morton undertook some poorly documented, preliminary experiments with ether inhalation. In an attempt at secrecy, he would obtain ether from different sources and try it out on one or two animals, and even on himself, as he was unable (or unwilling) to find other human volunteers. As he had no scientific training and was ignorant of even the basic chemistry of ether, he made little progress over the next few weeks. Morton knew that by finding a method of painless tooth extraction his fortune would be assured, and he was driven more by financial gain and less by the fame that might accrue from such a discovery.

In seeking out someone with knowledge about ether, Morton again visited his mentor from Medical School, Dr Jackson. At this meeting, Morton casually asked Jackson whether he thought ether inhalation could be used to extract teeth painlessly and safely. Jackson agreed it could and related how he himself had inhaled ether to overcome the unpleasant symptoms of having accidently inhaled chlorine gas a few years previously. He said he had even recommended its use to other colleagues. Perhaps the most important information Jackson emphasised at this meeting was that the ether used should be the highly purified sulphuric type and he recommended a supplier.

As Morton hurried away from this historic meeting, it is worth reviewing Jackson's own position at the time, bearing in mind subsequent events. Here was an eminent doctor and chemist who seemed to know that ether had pain-relieving properties and might be a suitable general anaesthetic. He had learned this from his own observations. He had given up medical practise some years ago but failed to recognise the potential of his own findings. Instead, he gave advice to Morton but was not interested enough to request to be present if and when the procedure was undertaken.

From knowledge gained at his meeting with Jackson, Morton now felt ready to use ether for tooth extraction. The very next day, 30 September 1846, he tested purified sulphuric ether on himself, inhaling its vapour from a handkerchief. He soon drifted into a state of unconsciousness, from which he quickly recovered. Perhaps the combination of his ignorance, inexperience and enthusiasm worked in Morton's favour, for, without further experimentation, he felt he was ready to try it out on a patient. That evening a man named Eben Frost came into his surgery in great pain and asked Morton to extract a tooth. Morton got Frost to inhale ether vapour from a handkerchief soaked in the solution. This rendered Frost unconscious and the troublesome tooth was extracted without pain.

Over the next few days Morton carried out further, mostly successful, extractions. He ensured that news of his discovery appeared in the local newspaper (Figure 3.8). Without anyone realising it, this must be one of the most important documents in medical history. It reads:

> *IMPORTANT DISCOVERY. For years surgeons and dentists have vainly sought for some means to alleviate pain, while severe operations were taking place. Many an individual has gone to his grave, through a dread of suffering, who might have been saved had he been willing to place himself in the hands of the surgeon; and persons have suffered for many a long year with their Teeth, rather than submit to have them extracted. It is acknowledged that extracting teeth is one of the most painful operations in surgery, but the pain being momentary is not so generally perceived. To remove all these difficulties, and allow even the most sensitive an opportunity for the performance of any surgical or dental operation, Dr Morton has discovered a compound to alleviate pain. He can, after administering it, extract a tooth, and the patient will not be sensible of the slightest pain. He has done it in repeated instances. He has made arrangements to extract Teeth, using his Compound, at his Rooms. No 19 Tremont Row.*

Morton was happy for colleagues to come and observe his methods. He also set to work to design a more convenient glass inhaler to deliver the ether vapour. Morton's thoughts quickly turned to how he could best benefit financially from his discovery. He realised there would be a huge market for painless tooth extractions,

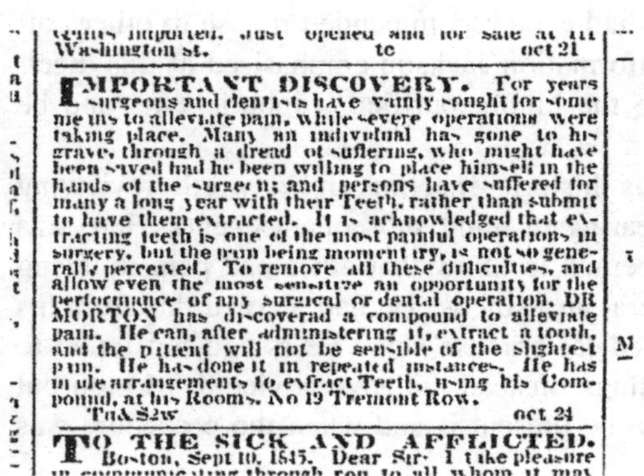

Washington st. tc oct 21

IMPORTANT DISCOVERY. For years surgeons and dentists have vainly sought for some means to alleviate pain, while severe operations were taking place. Many an individual has gone to his grave, through a dread of suffering, who might have been saved had he been willing to place himself in the hands of the surgeon; and persons have suffered for many a long year with their Teeth, rather than submit to have them extracted. It is acknowledged that extracting teeth is one of the most painful operations in surgery, but the pain being momentary, is not so generally perceived. To remove all these difficulties, and allow even the most sensitive an opportunity for the performance of any surgical or dental operation, DR MORTON has discoverad a compound to alleviate pain. He can, after administering it, extract a tooth, and the patient will not be sensible of the slightest pain. He has done it in repeated instances. He has made arrangements to extract Teeth, using his Compound, at his Rooms, No 19 Tremont Row.

TuA S2w oct 24

TO THE SICK AND AFFLICTED. Boston, Sept 10, 1845. Dear Sir- I take pleasure

Figure 3.8 Advert placed by William Morton in the *Boston Evening Transcript* (published as the *Daily Evening Transcript*), 31 October 1846, **17**(4991):4. *Source*: Courtesy of the U.S. Library of Congress.

but once the anaesthetic agent was known to be readily available ether, anyone would be free to use it.

Morton immediately sought to protect his financial interests by seeking to patent his discovery. He visited Robert H. Eddy, a patent lawyer, who deemed that the process of ether anaesthesia was patentable, although reading through it today, this seems extremely doubtful, as Morton was merely suggesting inhaling readily available ether. Eddy happened to be a friend of Jackson, and, on learning more about the discovery, felt that Jackson's name should be included. This would give Jackson recognition in the discovery, ensure that he would not oppose it if excluded and lend authority to the application. Morton agreed to this and a patent was drawn up on 27 October 1846, submitted to the patent office and approved a few months later as patent number 4848 (Figures 3.9 and 3.10). Regarding the financial implications, Eddy was to be awarded 25% of the profits for his help with the patent. Although named as a co-discoverer, Jackson was reluctant to be seen to benefit financially and, for a small consideration for his input (10%), he signed away his rights in favour of Morton, who claimed the major share of 65%. At this early stage, Jackson may still have been unsure as to the benefits of the discovery, while Morton was thinking mainly in terms of its use in dental extractions, having little awareness at the time of its wider application to general surgery.

UNITED STATES PATENT OFFICE.

C. T. JACKSON AND WM. T. G. MORTON, OF BOSTON, MASSACHUSETTS; SAID C. T. JACKSON ASSIGNOR TO WM. T. G. MORTON.

IMPROVEMENT IN SURGICAL OPERATIONS.

Specification forming part of Letters Patent No. 4,848, dated November 12, 1846.

To all whom it may concern:

Be it known that we, CHARLES T. JACKSON and WILLIAM T. G. MORTON, of Boston, in the county of Suffolk and State of Massachusetts, have invented or discovered a new and useful Improvement in Surgical Operations on Animals, whereby we are enabled to accomplish many, if not all, operations, such as are usually attended with more or less pain and suffering, without any or with very little pain to or muscular action of persons who undergo the same; and we do hereby declare that the following is a full and exact description of our said invention or discovery.

It is well known to chemists that when geon with the amount of ethereal vapor to be administered to persons for the accomplishment of the surgical operation or operations required in their respective cases. For the extraction of a tooth the individual may be thrown into the insensible state, generally speaking, only a few minutes. For the removal of a tumor or the performance of the amputation of a limb it is necessary to regulate the amount of vapor inhaled to the time required to complete the operation.

Various modes may be adopted for conveying the ethereal vapor into the lungs. A very simple one is to saturate a piece of cloth or sponge with sulphuric ether, and place it to

Figure 3.9 First page of patent for the use of ether as an anaesthetic in 1846.
Source: Courtesy of U.S. Patent Office.

2 **4,848**

phine—with the ether. This may be done by any way known to chemists by which a combination of ethereal and narcotic vapors may be produced.

After a person has been put into the state of insensibility, as above described, a surgical operation may be performed upon him without, so far as repeated experiments have proved, giving to him any apparent or real pain, or so little in comparison to that produced by the usual process of conducting surgical operations as to be scarcely noticeable. There is very nearly, if not entire, absence of all pain. Immediately or soon after the operation is completed a restoration of the patient to his usual feelings takes place without, generally speaking, his having been sensible of the performance of the operation.

From the experiments we have made we are led to prefer the vapors of sulphuric ether to those of muriatic or other kind of ether; but any such may be employed which will properly produce the state of insensibility without any injurious consequences to the patient.

We are fully aware that narcotics have been administered to patients undergoing surgical operations, and, as we believe, always by introducing them into the stomach. This we consider in no respect to embody our invention, as we operate through the lungs and air-passages, and the effects produced upon the patient are entirely or so far different as to render the one of very little while the other is of immense utility. The consequences of the change are very considerable, as an immense amount of human or animal suffering can be prevented by the application of our discovery.

What we claim as our invention is—

The hereinbefore-described means by which we are enabled to effect the above highly-important improvement in surgical operations—viz., by combining therewith the application of ether or the vapor thereof—substantially as above specified.

In testimony whereof we have hereto set our signatures this 27th day of October, A. D. 1846.

CHARLES T. JACKSON.
WM. T. G. MORTON.

Witnesses:
R. H. EDDY,
W. H. LEIGHTON.

Figure 3.10 Last page of patent for the use of ether showing list of signatories in 1846. *Source*: Courtesy of U.S. Patent Office.

In early October, within a few days of his first success, Morton approached the medical staff at the Massachusetts General Hospital, the same hospital where he had accompanied Wells in the unsuccessful trial demonstrating nitrous oxide anaesthesia. Dr Warren, the same surgeon involved in Wells's trial, was still interested enough to invite Morton to demonstrate his anaesthetic. The invitation from Warren may have been the result of efforts made by another staff member, a junior surgeon named Dr Henry Jacob Bigelow, who had attended a demonstration of ether at Morton's practice.

Warren was taking a considerable personal risk in allowing Morton to demonstrate his anaesthetic procedure without revealing its composition. This was against the ethics of Warren's own local medical committee, for should anything go wrong at the operation, Warren could land in serious trouble. As Morton was taking out a patent, it was assumed by Warren that the anaesthetic solution contained a unique substance(s) discovered by Morton. Morton referred to the anaesthetic solution as letheon, named after the Greek Goddess of sleep, Lethe. Although ether has a characteristic smell, Morton attempted to disguise it by adding additional chemicals.

Morton was invited to the operating theatre on Friday, 16 October, where a relatively minor operation was scheduled. In the presence of a number of staff and students, Morton anaesthetised the patient, 21-year-old Edward Gilbert Abbott, following which Dr Warren removed a tumour from the left side of the neck with no visible painful responses. Remembering Wells's previous unsuccessful attempt, Dr Warren remarked to the watching crowd that 'This is no humbug'.

The next day Morton was invited to repeat the performance. This second operation was performed by Dr George Haywood and involved removing a lipoma (a tumour composed of fat cells) from the arm of a woman. Again, the operation proved a success, with more people gradually realising its potential.

The first hint of future trouble to come occurred soon after the second operation when Dr Jackson met Dr Warren and claimed that he was the one who had instructed Morton to use ether, although crediting Morton with designing the inhaler.

To assess the full potential of his discovery, Morton was invited back on 7th November to participate in a much more extensive operation. However, Warren realised that he could undertake no further operations until he knew the composition of the anaesthetic agent and considered it safe for his patients. This posed a major problem for Morton for, once it was revealed that the anaesthetic solution was ether, the cat would be out of the bag. In this battle of wits between Warren and Morton, Morton 'blinked' first and, just before the operation was due to commence, disclosed the composition of his anaesthetic agent, although aware that this disclosure could prejudice his future financial goals. On hearing it was simply sulphuric ether, Warren agreed that the operation could take place.

The third operation required the amputation of a leg above the knee in a young female patient. Previously, an operation this severe was rarely carried out due to the high mortality rate associated with the shock, unbearable pain, prolonged bleeding and post-operative infection. Warren successfully amputated the limb and carefully sutured the wound without the need for the usual speed as the patient remained unconscious and motionless throughout the procedure. With this major demonstration of the power of his anaesthetic, Morton, a local dentist, initiated a new age in surgery, although it would be many years before inhalation anaesthesia was routinely used.

The first scientific paper describing the success of these initial operations under ether was written by Dr. Henry Bigelow and entitled 'Insensibility during surgical operations produced by inhalation' (Figure 3.11). It appeared in the November 18, 1846 edition of the *Boston Medical and Surgical Journal* and is a landmark in the history of medicine. It is not clear how Bigelow, a junior member of staff, had the audacity to report the work in which he played such a minor role ahead of the chief surgeon, Dr. Warren, who wrote his own paper on 9 December in the same journal (Figure 3.12). Both of these papers noted that ether was the subject of a patent.

Immediately following his great triumph, trouble was already brewing for Morton, as in the same 9 December edition of the journal, there was an article from a Dr J.F. Flagg who was unwilling to abide by what he regarded as an unworkable patent, bearing in mind the anaesthetic was simply sulphuric ether that was already in common usage.

Reports of these early operations under ether anaesthesia spread rapidly to Europe.

Professor Jacob Bigelow, Henry Bigelow's eminent father (who taught medicine and botany at Harvard), immediately upon learning the news about the discovery

THE

BOSTON MEDICAL AND SURGICAL JOURNAL.

VOL. XXXV. WEDNESDAY, NOVEMBER 18, 1846. No. 16.

INSENSIBILITY DURING SURGICAL OPERATIONS PRODUCED BY INHALATION.

Read before the Boston Society of Medical Improvement, Nov. 9th, 1846, an abstract having been previously read before the American Academy of Arts and Sciences, Nov. 3d, 1846.

By Henry Jacob Bigelow, M.D., one of the Surgeons of the Massachusetts General Hospital.

[Communicated for the Boston Medical and Surgical Journal.]

IT has long been an important problem in medical science to devise some method of mitigating the pain of surgical operations. An efficient agent for this purpose has at length been discovered. A patient has been rendered completely insensible during an amputation of the thigh, regaining consciousness after a short interval. Other severe operations have been performed without the knowledge of the patients. So remarkable an occurrence will, it is believed, render the following details relating to the history and character of the process, not uninteresting.

On the 16th of Oct., 1846, an operation was performed at the hospital, upon a patient who had inhaled a preparation administered by Dr. Morton.

Figure 3.11 The introductory page to the first paper in a recognised scientific journal with the news of the discovery of general anaesthesia by Dr H.J. Bigelow on 18 November 1846. It was entitled 'Insensibility during surgical operations produced by inhalation'. *Boston Medicine and Surgery Journal,* **35**:309–317.

of ether anaesthesia sent a letter to his lifelong friend, Dr Francis Boott, in London. This letter was mailed on 3 December and arrived in London 2 weeks later. Dr. Boott passed on the news to his friend and neighbour, a dentist named James Robertson, and to the surgeon Dr Robert Liston. Robertson painlessly extracted a tooth using ether on 19 December, while Liston carried out the first major operation in England under ether anaesthesia when he amputated a leg at University College Hospital on 21 December. Boott then had a copy of Henry Bigelow's paper published in *The Lancet* on 2 January 1847.

Like nitrous oxide, ether did not always produce successful anaesthesia, and it took many years before the proper procedures for delivering enough of the vapour were determined, the correct temperature being an important factor. Much of the early work on improving the understanding of ether anaesthesia and the techniques required was made by a London surgeon, Dr John Snow. Between 1848 and 1858, he administered nearly 4,500 anaesthetics for both surgeons and dentists and became the first specialist in this new field. In addition to this major contribution

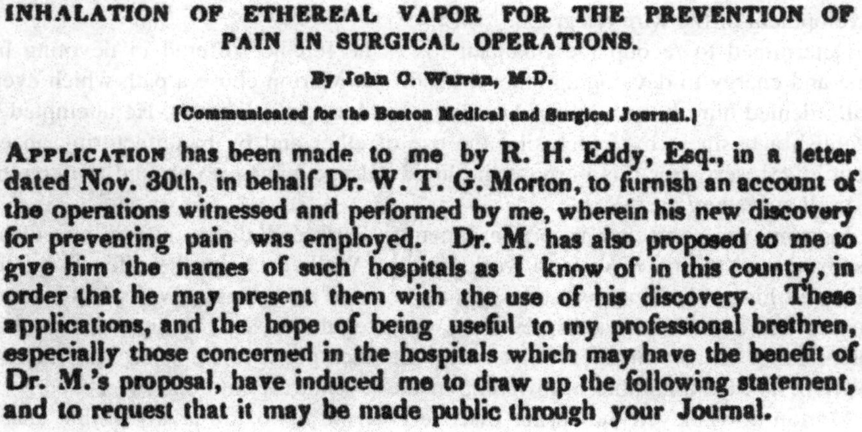

INHALATION OF ETHEREAL VAPOR FOR THE PREVENTION OF PAIN IN SURGICAL OPERATIONS.

By John C. Warren, M.D.

[Communicated for the Boston Medical and Surgical Journal.]

Application has been made to me by R. H. Eddy, Esq., in a letter dated Nov. 30th, in behalf Dr. W. T. G. Morton, to furnish an account of the operations witnessed and performed by me, wherein his new discovery for preventing pain was employed. Dr. M. has also proposed to me to give him the names of such hospitals as I know of in this country, in order that he may present them with the use of his discovery. These applications, and the hope of being useful to my professional brethren, especially those concerned in the hospitals which may have the benefit of Dr. M.'s proposal, have induced me to draw up the following statement, and to request that it may be made public through your Journal.

Figure 3.12 The first paragraph in the second paper in a recognised scientific journal with the news of the discovery of general anaesthesia written by Dr J.C. Warren. He was the surgeon most closely associated with its introduction in 1846. It was entitled 'Inhalation of ethereal vapour for the prevention of pain in surgical operations' and appeared in the *Boston Medicine and Surgery Journal,* 9 December, **35**:376–379.

to medicine, Dr Snow was also responsible for an even greater contribution by discovering that cholera, a fearful disease of plague-like proportions, was a waterborne illness spread by polluted drinking water.

The Fight for Recognition

Following his successful demonstration of surgical anaesthesia, there seemed little more that Morton needed to do. With no proper education and being a 'mere' dentist, he had stumbled upon one of the greatest discoveries in medicine. He would surely be lauded by everyone as a great humanitarian. He subsequently received many accolades, including the award of a medical degree. Morton could have shared his success with Jackson, who had provided help along the way, even though Jackson appeared disinterested in the progress of Morton's trials and had not been present at any operation. But he chose not to.

Relations between Morton and Jackson were not helped by the fact that, unbeknown to Morton, on 13 November 1846, Jackson had sent a letter to the Paris Academy of Science, to whom he was already known. He sent this letter immediately following the first successful operations of ether anaesthesia in Massachusetts General Hospital. In it Jackson claimed to be the true discoverer of the pain-relieving properties of ether and that he had persuaded a dentist to administer it.

His behaviour is reminiscent of his previous stance with regard to Morse and the development of the wire telegraph.

Determined to recoup the financial losses he felt he suffered in devoting his time and energy to developing ether anaesthesia, Morton chose a path which eventually denied him the rewards of his discovery during his lifetime. He attempted to establish a business by franchising the use of ether and by manufacturing anaesthetic dispensers. For this purpose, he hired salespeople to go around the country and sell anaesthetic services.

Morton even contacted his original mentor, Horace Wells, to inform him of his discovery and to invite Wells to work for him. Wells declined the offer. Since the failure of his public demonstration of nitrous oxide anaesthesia, Wells had become morose and ceased to practise dentistry on a regular basis, although occasionally volunteering to administer nitrous oxide. In attempting to improve his financial position, he tried a number of different business ventures, but none succeeded.

Morton downplayed the earlier discovery of nitrous oxide as a possible anaesthetic agent, and this prompted Wells to reestablish his own priority in the discovery of inhalation anaesthesia. He travelled to Paris at the beginning of 1847, where he was feted and acknowledged by the French as the discoverer of pain-free surgery. Returning to America after this 3-month trip, at the end of March 1847, Wells published a major pamphlet on anaesthesia, setting out the background to his discovery and establishing his precedence in the field. Whatever the particular general anaesthetic used, whether nitrous oxide or ether, it worked on the same basic principle he discovered, namely that an exhilarating agent when inhaled could cause insensibility to pain. This defence was also published in *The Lancet* of May 1847.

Wells finally moved to New York, where he still occasionally administered nitrous oxide anaesthesia. He continued to experiment with anaesthetic agents, administering gases to himself, including chloroform, which had been introduced as a new anaesthetic in 1847 by Dr James Young Simpson, a professor of obstetrics in Edinburgh. Simpson had been in London at the end of 1846 and heard the news of Liston's success with ether. He realised ether's potential for use in pain control during childbirth and was the first to use and report on it in this situation. His search for other volatile agents capable of producing anaesthesia but without some of the drawbacks of ether soon led to him using chloroform. It was Simpson who arranged for Queen Victoria to inhale chloroform during the birth of her son, Prince Leopold. As chloroform, like other anaesthetics of the time, did result in some fatalities, it is fortunate for the history of anaesthesia that she suffered no ill effects, but on the contrary was delighted with the pain relief. Simpson had to overcome opposition from various quarters, including the church, where some people believed pain relief during childbirth was wrong, as the Bible said (Genesis 3:16): 'in sorrow thou shalt bring forth children'.

Wells probably became addicted to sniffing chloroform, which would account for the rapid deterioration in his mental state, culminating with him being imprisoned for throwing acid over two women. While in prison, on 24 January 1848, he committed suicide by severing an artery after inhaling chloroform. He was just

33 year old. His estate was declared insolvent, and his wife was forced to sell the contents of their house and dental practice to pay off his debts.

It was only some time after his death that the significance of Wells's contribution to surgery was fully realised. In 1875 a bronze statue of Wells was erected in Hartford, Connecticut. At the base of the statue his words concerning the use of nitrous oxide are quoted: 'I was desirous that it should be as free as the air we breathe'.

With the use of ether established as an anaesthetic for general surgery, Morton now realised that he had grossly underestimated the task of protecting his patent for the use of ether across America, and it was not long before he was in financial difficulties. His solution to this setback was to petition the U.S. Congress, claiming that ether was being used in hospitals in breach of his patent. For a substantial lump sum, he wished to be compensated for this and the loss of time and money in developing anaesthesia, which had provided great benefit to the country. Although government establishments were breaking his patent every day, Morton did not wish to take them (or every medical and dental practitioner) to court if this could be avoided, due to the time and cost involved and the uncertainty of the outcome.

This decision to approach Congress was to dominate the rest of Morton's life. He unsuccessfully petitioned on four separate occasions. A major impediment to his petition was his claim that the discovery of ether anaesthesia was his alone. This was rejected by Jackson, who argued that the discovery of general anaesthesia was his and that Morton had only been following his instructions. The events in Congress were further complicated by the reemergence of Horace Wells's earlier work with nitrous oxide. His widow put forward a counterclaim on behalf of her son that her husband was the true discoverer of anaesthesia and, in providing such a benefit for humanity, had not received, and did not wish for, any financial reward but had died and left his family destitute.

The effort to gain the recognition he felt he deserved may have taken its toll on Morton's health. In New York on 15 July 1868, he died just short of his 49th birthday. He was buried in Mount Auburn cemetery, Cambridge, Massachusetts. A monument was later erected over his coffin and inscribed with the words 'W.T.G. Morton, Inventor and Revealer of Anaesthetic Inhalation'.

With the untimely deaths of both Wells and Morton, the life of the third person involved in the controversy surrounding the discovery of anaesthesia also came to a sad end. Jackson died in 1880 at age 68, in an asylum for the insane, where he had spent the last 7 years of his life.

Following the death of Wells in 1848, nitrous oxide was used only occasionally as an anaesthetic agent for tooth extraction. 'Professor' Colton, who had introduced Wells to laughing gas, settled in New York in the 1860s. Here he established Colton Dental Associates, whose principal purpose was to reintroduce nitrous oxide anaesthesia as a means of painless tooth extraction. It was Colton who was chiefly responsible for the revival and spread of this technique throughout America and Europe. Improvements followed after 1868 when nitrous oxide was combined with oxygen by the American surgeon Dr Edmund Andrews. This provided a safer and more prolonged anaesthesia. Bearing in mind the millions of people requiring tooth extraction, the benefits achieved by the discovery of nitrous oxide are incalculable.

Figure 3.13 U.S. postage stamp for 1940, honouring Dr Crawford Long.

Having described the crucial and beneficial role that dentists played in the discovery of anaesthesia, it must also be acknowledged that it was a French dentist in 1847 who was the first person found guilty of assault on not one, but two, young women while they were under the influence of ether for the purpose of tooth extraction. The practitioner, Dr Laine, protested his innocence but was sentenced to 6 years' hard labour. An article relating to the affair appeared in *The Times* on 5 November 1847.

Dr Crawford Williamson Long (1815–1878)

In any history concerning the discovery of general anaesthesia, mention needs to be made of Dr Crawford Long. A newly qualified medical graduate of the University of Pennsylvania, he was a country doctor in the small town of Jefferson, Georgia. There, in 1842, he carried out a small number of pain-free, surgical operations using ether anaesthesia. Gaining no encouragement from local doctors and being so isolated, he soon ceased the procedure. He only published his findings belatedly in 1849 after learning of Morton's discovery. Because of this, he is not considered in the same light as Wells and Morton, who publicised their findings as soon as and as widely as they could. The U.S. Postal Service saw fit to issue a commemorative stamp in 1940, honouring Long (Figure 3.13), but it has never issued one honouring either of the two dentists, Wells and Morton.

4 What is a Tooth?

Metaphors using the word tooth have entered our language, as in phrases like tooth and nail, long in the tooth, in the teeth of, by the skin of one's teeth, to cut one's teeth on and a kick in the teeth. The saying 'Don't look a gift horse in the mouth' also has dental connotations. You can accurately tell the age of a horse by looking at its teeth, first by determining what teeth are present (milk teeth or permanent) and then by the degree of wear of its front permanent teeth (on the grinding edges of each one is a depression or mark which gets worn away at known intervals). The actual meaning of this phrase is, if someone wants to give you something for nothing, don't start examining it too closely.

Although everyone knows what a tooth is, it is not easy to formulate a precise definition. Referring to human teeth, most people would say:

1. They are hard and exposed in the mouth (the crown part) where their main function is to cut up or crush food prior to swallowing.
2. They have different shapes, with grinding teeth at the back (molars) and more simple cutting teeth in the front (incisors).
3. They have roots embedded within the jaws.
4. There are only two sets.

Teeth are found in all the major groups of backboned animals (vertebrates). In ascending order of complexity, these groups are:

- Fish
- Amphibians (animals that lay their eggs in water)
- Reptiles (animals that lay their eggs on land, as they have hard protective shells)
- Mammals (animals that suckle their young)

Birds are excluded because they do not have teeth (but see Chapter 11).

The teeth in fish, amphibians and reptiles differ from those in mammals in four fundamental ways (Figure 4.1):

1. They consist only of crowns that are attached to the surface of the jaw.
2. They all are of a similar, simple shape (usually triangular or cone-shaped), whereas in most mammalian jaws, the teeth are more complex and have different shapes in different parts of the jaw (incisors, canines, premolars and molars).
3. They are continuously replaced throughout life, so there are far more than just the two sets seen in mammals (see Chapter 9).

Nothing but the Tooth. DOI: http://dx.doi.org/10.1016/B978-0-12-397190-6.00004-3

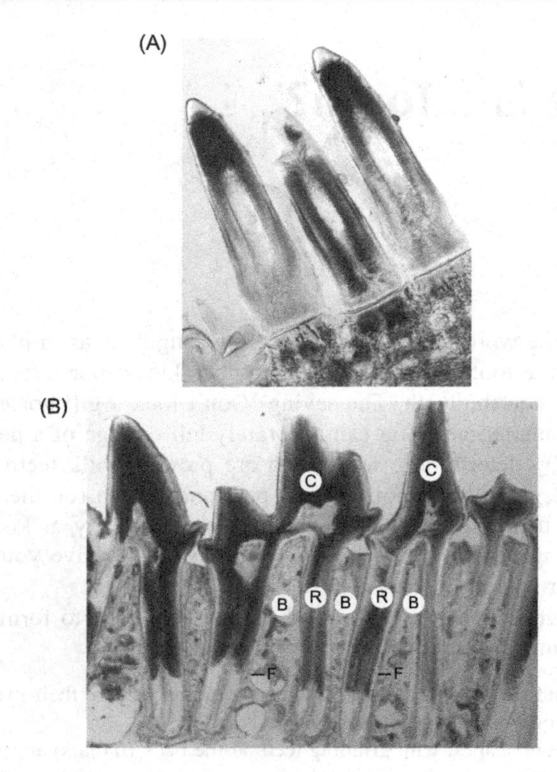

Figure 4.1 (A) Section showing three teeth in the conger eel that are attached at their bases to the jawbone and supported without the presence of roots. (B) Section through the jaw of a mammal (loris from Madagascar) showing parts of five teeth, all having a crown (C) in the mouth supported by a root (R) attached to a bony socket (B) in the jaw by a fibrous joint (F). *Source*: Courtesy of the Hunterian Museum at the Royal College of Surgeons.

4. Their simple-shaped teeth are not capable of coming together to cut or grind up the food. The teeth are used to prevent the escaping of the prey, which is then swallowed head first. Mammalian teeth are able to come together (occlude) and crush/cut the food into smaller pieces prior to swallowing.

All true teeth have the same structure based on a hard core of dentine that surrounds and protects the central, soft, sensitive, dental pulp. The dentine is covered on the crown by an even harder layer, the enamel. The extra component in mammalian teeth, the root, has a core of dentine covered by a very thin layer of cement (Figure 4.2A), allowing the tooth to be anchored to the surrounding bone by a fibrous joint (Figure 4.1B, label F).

The hardness and resistance to wear of dentine and enamel are related to the presence of mineral crystals of calcium phosphate (and some calcium carbonate). Both enamel and dentine have a complex structure, with dentine being permeated by minute tubes (Figure 4.2B) and the enamel showing many regular lines (due to sudden

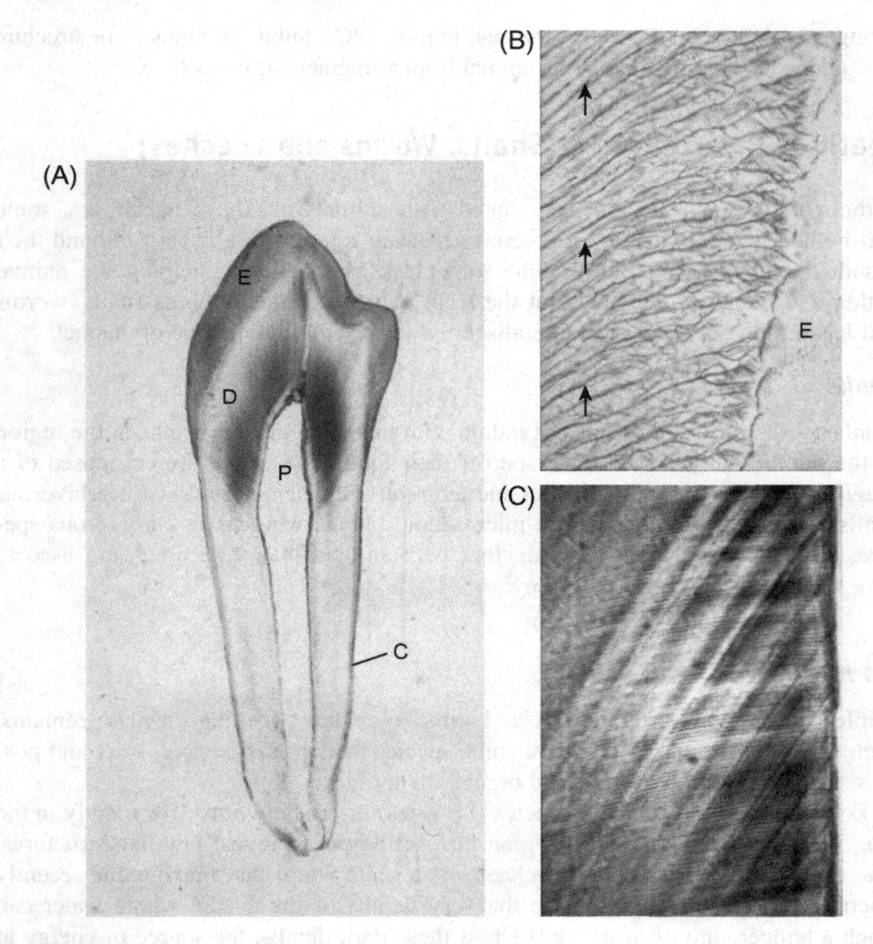

Figure 4.2 (A) Section through a human tooth showing the central pulp cavity (P) (empty due to the method of preparation), protected by dentine (D) forming the body of the tooth. In the crown, the dentine is covered by enamel (E). In the root, the dentine is covered by a thin layer of cement (C), which allows it to be attached to the bone of the socket by a fibrous joint (periodontal ligament). (B) High-power view of the dentine. The arrows indicate the numerous tubules that characterise this tissue. E = overlying enamel. (C) High-power view of the enamel. The complex arrangement of its crystals forms a large number of characteristic lines. The horizontal lines represent enamel rods or prisms, and the oblique lines represent weekly incremental lines (see Chapter 13).
Source: From B.K.B. Berkovitz, G.R. Holland and B.J. Moxham, 2009. *Oral Anatomy, Histology and Embryology*. 4th edition. Elsevier.

changes in the orientation of its crystals; Figure 4.2C). Subtle differences in structure may allow an expert to identify an animal from a fragment of its tooth.

Teeth of Invertebrates (Snails, Worms and Leeches)

Although true teeth are only associated with animals that have backbones, some soft-bodied invertebrates have structures that are referred to as 'teeth' around their mouth openings. They perform the same task as true teeth, helping the animal gather and break up the food, but the teeth of invertebrates such as snails, worms and leeches do not have any mineralised tissues resembling dentine or enamel.

Snails

Snails possess a structure called a radula with hundreds of tiny 'teeth' in the region of the mouth that are used to scrape off their food. The 'teeth' are composed of a toughened, horny, organic material and are replaced when worn down. Herbivorous snails use the radula to graze on microscopic plants, whereas in carnivorous species, the teeth are sharper and, together with an acid that is secreted, are used to bore through the shells of their prey.

Worms

While worms may be thought of as harmless creatures feeding on plant remains, bacteria, fungi and algae, there are some species that are serious predators and possess 'teeth' composed of hardened organic material.

Polychaetes (poly = many, chaetes = bristles) or bristle worms live mainly in the sea. Some have mouths that they can turn inside-out to reveal jaw-like structures. One of the most remarkable polychaetes is a scale worm that inhabits the recently discovered hydrothermal vents in the very depths of the ocean, where water can reach a temperature of nearly 400°C! At these dark depths, the source of energy at the base of the food chain does not come from plants via the sun (photosynthesis) but from the oxidation by bacteria of inorganic molecules such as hydrogen sulphide (chemosynthesis). The mouth of this scale worm is surrounded by a number of joint-like appendages that have numerous 'teeth' to help them grasp and break up the bacteria and small organisms on which the worm feeds (Figure 4.3).

Another polychaete with a fearsome reputation is the bobbit worm. About 2 cm in width, it has been reported to reach a length of 3 m! It burrows in the ocean floor, lying in wait for its prey. Around its mouth, it has up to four pairs of jaw-like claws with 'teeth' (Figure 4.4). It strikes with lightening speed and preys on small fish.

Leeches

There are many species of leeches, each preying on different hosts. The medicinal leech, widely used in the eighteenth and nineteenth centuries for human blood-letting, is now being reintroduced into hospitals to relieve bruising following plastic

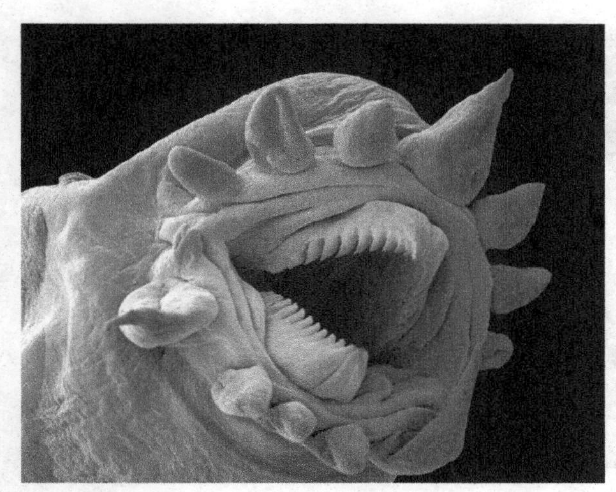

Figure 4.3 Mouth parts of a polychaete (scale worm) from a deep ocean hydrothermic vent with tooth-like projections. The mouth is surrounded by a number of sensory antennae. *Source*: Courtesy of SPL/ Barcroft.

Figure 4.4 Mouth parts of the bobbit worm containing tooth-like projections. *Source*: ©Ethan Daniels/ SeaPics.com.

surgery. One specialist in Germany places leeches around painful, osteoarthritic knee joints and claims they can provide pain relief lasting for several months.

The leech unusually has three jaws, each surmounted by up to 100 tiny 'teeth' (Figures 4.5 and 4.6). Its bite, which is painless, forms a Y-shaped wound (Figure 4.7) through which the leech sucks blood. Leech saliva contains a cocktail of important molecules. One acts as a local anaesthetic, another widens nearby small blood vessels and another (hirudin) acts as an anticoagulant. With these additives, a leech can drink about 10–15 cc of blood before releasing itself, while a further 20–50 cc of blood is lost from the wound prior to a blood clot being formed. The outline of a leech bite made by the three jaws is reminiscent of the Mercedes car logo (compare Figures 4.7 and 4.8).

Unlike snails and polychaetes, leech 'teeth' have been found to contain mineral salts of calcium, like true teeth, but little is known about their structure.

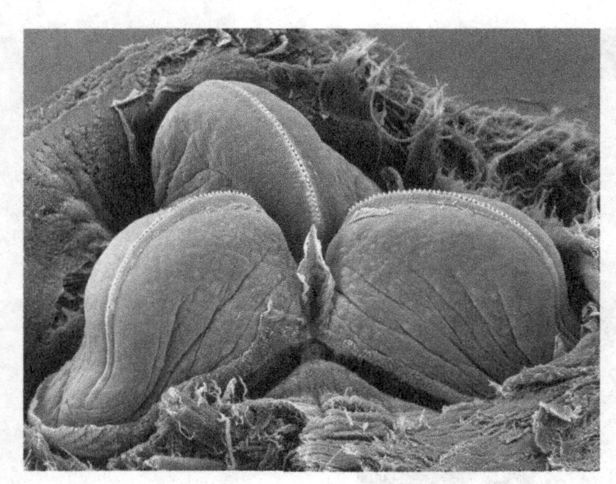

Figure 4.5 Image showing the three jaws of the leech, each surmounted by many small teeth.
Source: Courtesy of Eye of Science/Science Photo Library.

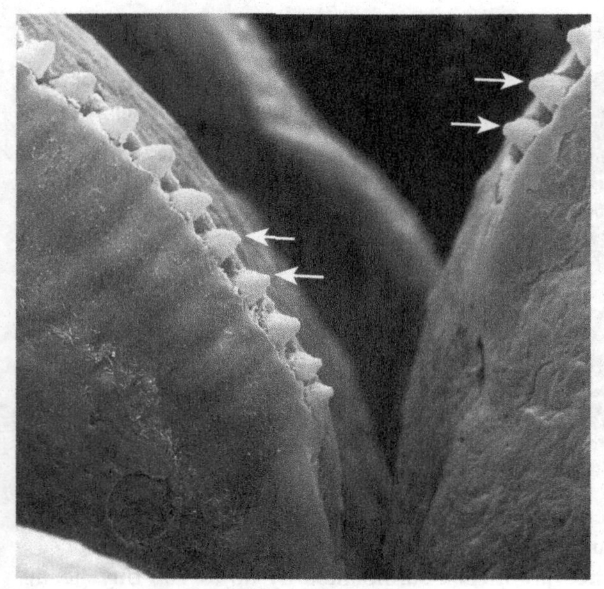

Figure 4.6 High-power view of the jaws of a leech showing surmounted by teeth (arrows).
Source: Courtesy of Eye of Science/Science Photo Library.

Teeth of Vertebrates (Fish, Amphibians, Reptiles and Mammals)

Lampreys

The most primitive of the fishes are lampreys and hagfishes. These do not have jaws around their mouths and hence are termed agnathans (a = without, gnathos = jaw). Like sharks and rays, they have a soft skeleton made of cartilage that does not fossilise. The sea lamprey is an eel-like parasite with numerous sharp,

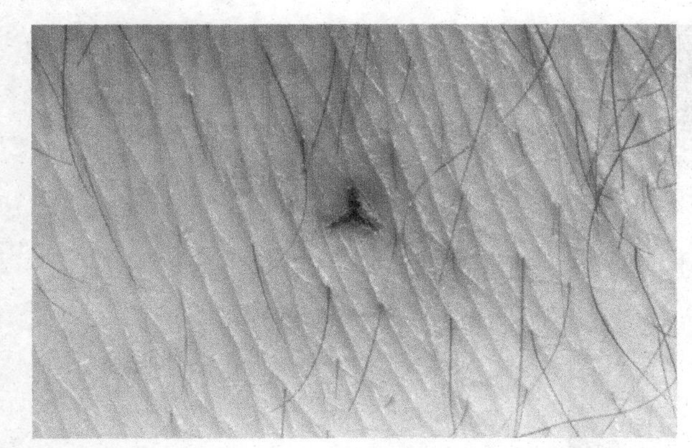

Figure 4.7 Bite from a leech showing a Y-configuration. *Source*: Courtesy of Scientifica Visuals Unlimited/Science Photo Library.

Figure 4.8 The Mercedes car logo showing a resemblance to the outline of a leech bite seen in Figure 4.7.

hard, hollow teeth. It uses the sucker around its mouth to attach itself to other fish. With its many teeth around the edges of the sucker, it is extremely difficult to dislodge. It rasps away the flesh of the host with its tongue, which also has ‘teeth’ on it (Figure 4.9). There may be up to 120 teeth present. They do not have the structure or mineral content of true teeth, being composed of a tough, horny protein found in other components of the skin, such as claws and fingernails (and also the horn of a rhinoceros). The teeth are continuously replaced. The saliva of the lamprey has an anticoagulant to maintain blood flow while it feeds.

Fish, Amphibians and Reptiles

Fish

As an example of a fish dentition, the barracuda is illustrated in Figure 4.10. It typifies the teeth of the lower vertebrates, with numerous teeth having the same basic simple shape (although differing in size) distributed on many bones

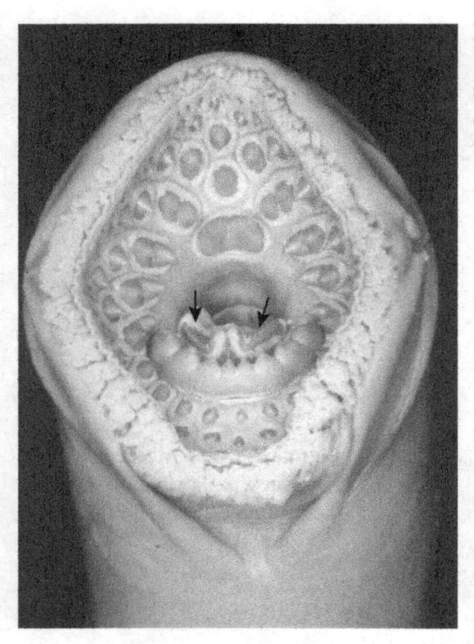

Figure 4.9 Sucker-like mouth of a sea lamprey. The teeth are arranged in whorls around the mouth opening. The lamprey rasps off flesh from its host with teeth that lie on the tongue (arrows).

Figure 4.10 Skull of barracuda showing rows of different-sized, cone-shaped, pointed teeth. *Source*: Courtesy of the Hunterian Museum at the Royal College of Surgeons.

surrounding the mouth. Their size and sharpness would indicate a fierce predator eating other fish. In some fish, such as trout, true teeth are even present on the tongue. While the dentine of the teeth may be similar to that of mammals, their enamel is simpler in structure and is very thin.

Amphibians

As an example of the teeth in amphibians, frogs have a row of very small teeth in the upper jaw, and teeth are absent in the lower jaw. Toads have no teeth at all. Frogs and toads grab their food using their large, sticky tongues.

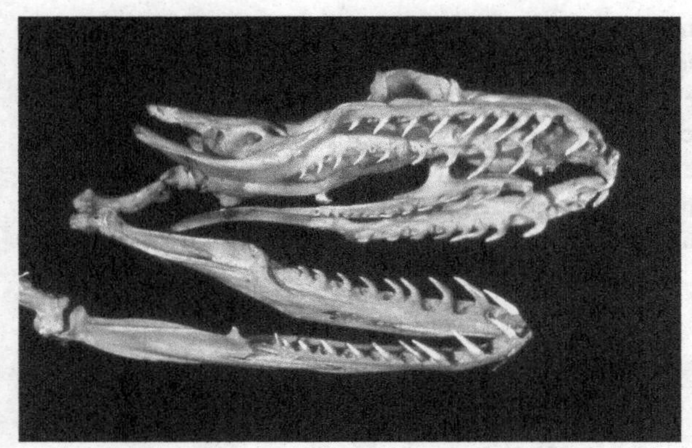

Figure 4.11 Skull of a python. The upper jaw (upper part of image) shows two rows of narrow, pointed, recurved teeth on each side, while the lower jaw (lower part of image) has a single row in each half.

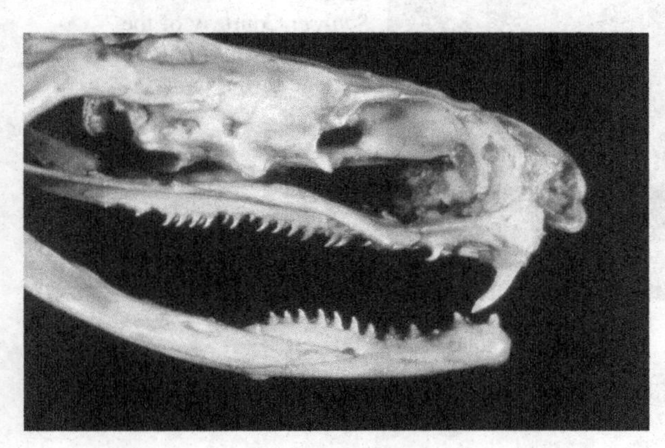

Figure 4.12 Skull of an adder (*Vipera berus*). The enlarged fang lies at the front of the upper jaw. The poison is injected directly into the prey through a channel within the tooth, like a hypodermic needle.

Reptiles

Taking snakes as an example of reptile dentitions, their prey may be killed by muscular constriction before swallowing (as seen in pythons), swallowed while alive, which can even be other snakes, or poisoned.

The python has two upper rows and one lower row of teeth in each half of the jaw (Figure 4.11). The teeth are curved backwards and help guide the swallowed prey in the right direction. The bones of the skull are very flexible at their joints and allow snakes to swallow prey larger than the diameter of the snake itself.

In poisonous snakes, the poison serves three functions: to immobilise/kill the prey, to start the digestive process and for defense. A tooth on each side of the upper jaw has enlarged into a fang, and there may be a variable number of other smaller teeth that help to grip and swallow the prey. When the snake strikes, poison passes via the fangs into its prey. The poison is produced in a specially modified salivary gland and contains a cocktail of molecules, some of which start the process of digestion (see also Chapter 6, page 90). In some snakes, the fang lies at the front of the jaw (front-fanged snakes; Figure 4.12), while in others, it lies further back (rear-fanged snakes). The position of the fang has functional significance in that front-fanged snakes are the most venomous. Here, the poison is injected directly into the prey through a channel contained within the fang itself (Figures 4.13 and 4.14). The hollow fang is like a hypodermic needle. Once it has struck, the snake disengages and leaves its prey to die before swallowing it whole. Rear-fanged snakes are less poisonous. The poison is not injected directly into the prey but runs down a groove on the tooth, and the snake needs to chew the victim to enable it to get enough poison into it. Any remaining teeth in the jaws are used to hold the prey and direct it backwards.

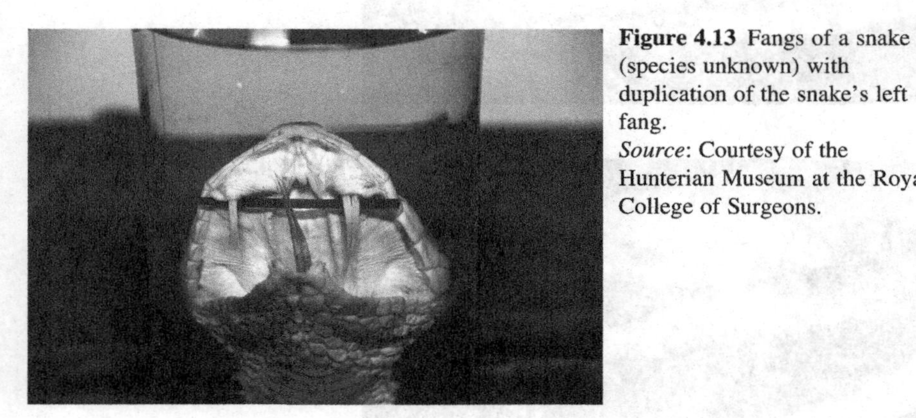

Figure 4.13 Fangs of a snake (species unknown) with duplication of the snake's left fang.
Source: Courtesy of the Hunterian Museum at the Royal College of Surgeons.

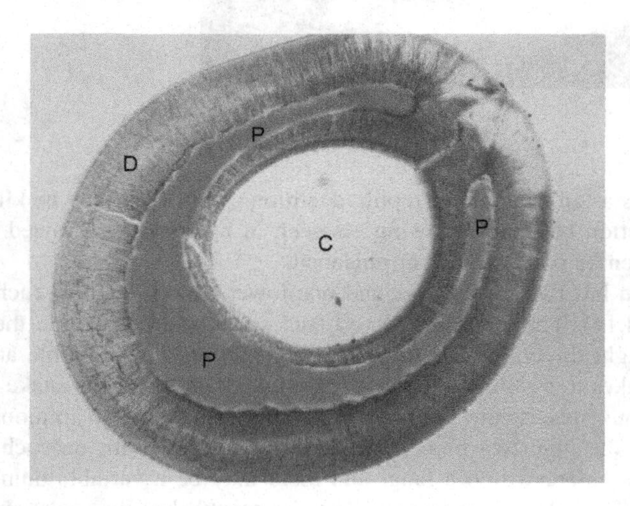

Figure 4.14 Cross section of the poisonous fang of a venomous bushmaster snake (*Lachesis muta*) from Central America. The normal central pulp chamber (P) is displaced to one side as a separate central chamber (C) develops to provide a channel that delivers all the poison without any waste into its victim. The rest of the tooth is composed of dentine (D), whose tubular structure is evident.
Source: Courtesy of the Hunterian Museum at the Royal College of Surgeons.

The fangs normally rest against the roof of the mouth and only become erected when the snake strikes its prey. As an example, rattlesnakes have a number of replacing fangs in differing stages of development. There are two rows on each side, which are shed alternately and so it is not unusual for a snake to have two fangs one side for a short period (Figure 4.13). The teeth in snakes may be shed every month.

While dealing with prey, teeth may break off and may be swallowed with the prey. In these situations, the teeth may pass through the intestinal tract and be found in the faeces.

One type of snake that breaks all the rules as far as teeth are concerned is the egg-eating snake. This snake has no teeth at all. As its name indicates, it eats eggs. It swallows them whole, even though the egg may have a far greater diameter than that of the snake. Once swallowed, the egg is compressed against a hard, bony projection from one of its vertebrae, which cracks the egg. All the nourishing liquid from within the egg is absorbed, after which the snake regurgitates the shell.

Alligators and crocodiles are the exception to the rule with regard to non-mammalian vertebrates in that their teeth have roots which, as in mammals, are attached to a bony socket by a fibrous joint (the periodontal ligament) (Figures 4.15 and 4.16). This allows them to generate large biting forces.

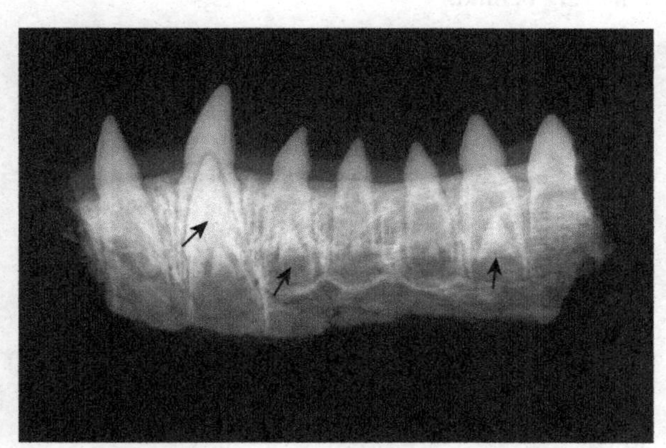

Figure 4.15 X-ray of part of the jaw of a caiman, a small type of crocodile found in Central and South America. The teeth have roots embedded in the jaw. Beneath most of the teeth, developing replacing teeth can be seen at different stages of development (arrows).

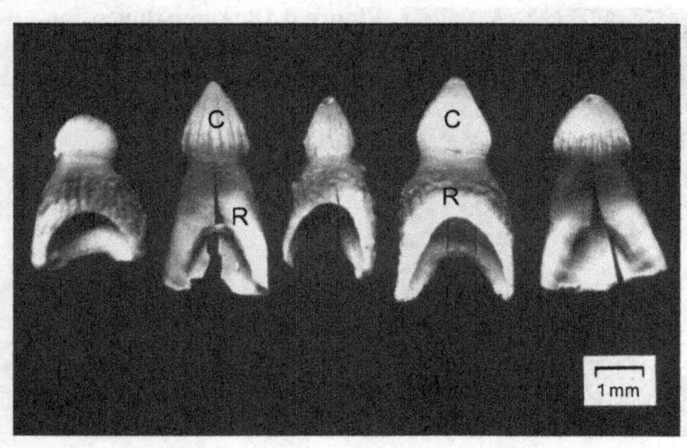

Figure 4.16 Caiman teeth showing the presence of both crown (C) and root (R) in the teeth of this crocodilian reptile. The roots have been eaten away at the base by underlying developing replacing teeth.

A softer, caring side of these great lumbering creatures can be seen when they carry their young in their powerful jaws down to the river. How can they carry out this delicate action without damaging the young? The answer is that there are sensitive pressure receptors distributed in the mouth and particularly in the fibrous joints supporting the roots of the teeth. These pressure receptors reflexly control and limit the pressure between the teeth.

As discussed in Chapter 9, page 120, each tooth in the Nile crocodile is continuously replaced throughout its lifetime, so an X-ray of the jaw of these and related creatures will always show the presence of replacing teeth (Figure 4.15).

Reptiles conquered the land environment by evolving a hard shell in the protection of which their young could develop. This was only possible if, at the same time, a mechanism existed to allow the young to break out of the egg. This difficulty was overcome with the aid of a specialised structure at the front of the beak or upper jaw, which is lost soon after hatching. In many reptiles (such as turtles and crocodiles) and all birds, this feature is known as the caruncle, a horny, tough, thickening of skin. However, in some lizards (e.g. the green lizard) and snakes (e.g. the python), a true tooth, the egg tooth, is present. It is a true tooth as it possesses a core of dentine (Figures 4.17 and 4.18). The egg tooth develops early and is larger than the normal teeth lying behind.

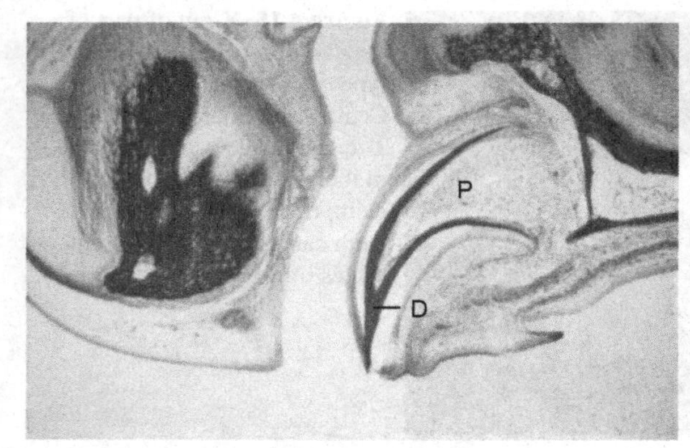

Figure 4.17 Stained microscope slide showing the developing egg tooth at the front of the upper jaw in the rainbow lizard. It is made up of stained dentine (D) surrounding the central dental pulp (P).
Source: Courtesy of Dr J.S. Cooper.

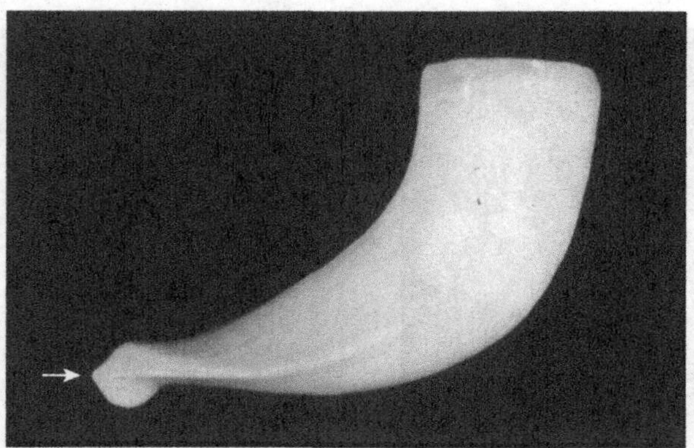

Figure 4.18 Three-dimension plastic model showing the egg tooth of the rainbow lizard from the side. Arrow indicates its tip.
Source: Courtesy of Dr J.S. Cooper.

Some primitive reptiles, such as geckos, have two egg teeth of equal size. In certain lizards, two egg teeth are initiated, one on each side, but only the right one enlarges. In the python and corn snake, although two egg teeth develop in the midline, they fuse together to form a single egg tooth.

Mammals

In mammals, teeth are used both for grasping and breaking up food and have a variety of shapes within the same jaw. They show adaptations to deal with three main types of food: plants and grasses (herbivores), meat (carnivores) and insects and worms (insectivores). Because of this, it is usually easy to tell from an animal skull the type of diet the animal would have eaten. Animals such as pigs and humans are termed omnivores, indicating a more varied diet. Their teeth are less specialised and the molars have simple, rounded cusps. Except for some of the toothed whales, no mammal has more than 44 teeth in its jaws (humans and great apes having 32), and the teeth are limited to single rows in each jaw.

Herbivores

Herbivores have teeth with broad, grinding, roughened surfaces as exemplified by sheep, cows and horses (Figure 4.19). As they are subjected to constant abrasive wear, the teeth tend to have very large grinding surfaces, long crowns and short roots. The jaws allow for lots of sideways movements during chewing that are necessary to crush and break up the plant material. As there is little nourishment in this type of food, many hours a day are occupied in chewing enough food. Herbivores also have big stomachs and can chew the cud (i.e. regurgitate its stomach contents into its mouth for chewing a second time). The next time you are at a zoo, just stand in front of a camel while it is chewing. It has a regular cycle, with one chew on the left alternating with one on the right, the lower jaw tracing out a perfect figure of eight.

Carnivores

Carnivores have to actively catch their prey and many are built for speed. They have very prominent, tusk-like canine teeth that are used as weapons to throttle and suffocate their prey. In the molar region, unlike the broad crushing teeth of

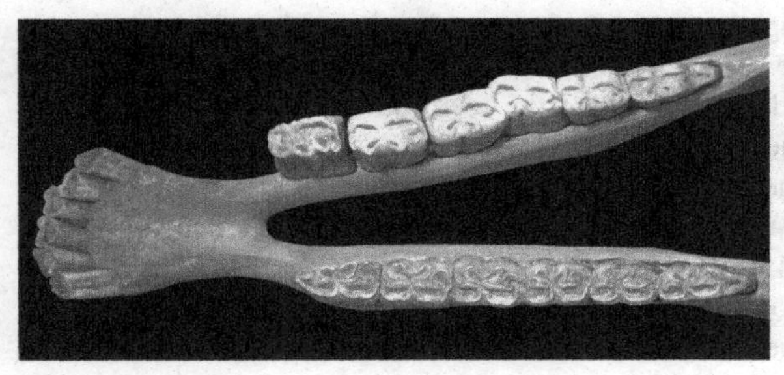

Figure 4.19 The lower jaw of a horse. Note the broad, roughened grinding surface of its battery of cheek teeth. *Source*: Courtesy of the Hunterian Museum at the Royal College of Surgeons. Photographed by M. Farrell.

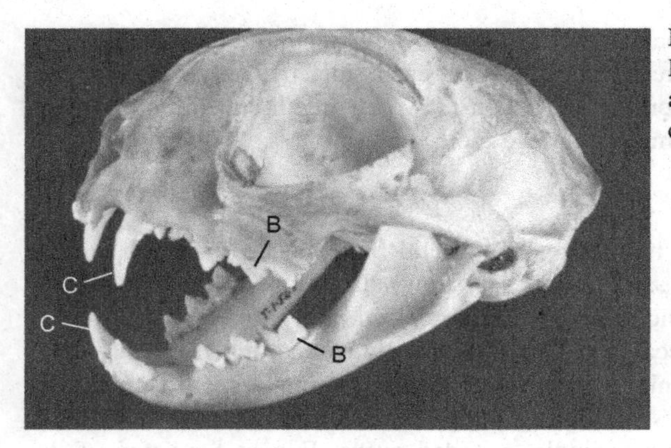

Figure 4.20 The skull of a cat. Note the enlarged canine (C) and the elongated blade-like cutting cheek teeth (B).

herbivores, the teeth are compressed into sharp blades, which slice meat off the bone like a pair of scissors (Figure 4.20). This type of slicing action requires only a hinge-like up and down movement, so that carnivores cannot move their jaws sideways, a movement that would be impossible anyway due to the interlocking large canines. In the less-specialised carnivorous dentitions, such as in the dog family, there is an additional broad molar tooth at the back of the jaws for use in crushing bones. As meat is very nutritious, carnivores do not need to feed as often as herbivores and do not have large stomachs.

The canines of carnivores reached their zenith in the extinct sabre-toothed cat (*Smilodon*) that lived between 2.5 million and 500,000 years ago. Its upper canine teeth reached a length of nearly 30 cm, leaving no room for a large lower canine (Figure 4.21). The sheer size of the canines can make it seem that eating or attacking prey would be a problem. However, sabre-toothed cats had the ability to open their mouths very wide (up to 120° as opposed to about 70° in large, living carnivores such as the panther) to allow for the extreme length of the upper canine. Its skull was not designed to handle the pressure of biting through bone but was designed to allow the canines to 'go for the jugular' and to rip off big chunks of soft flesh.

Insectivores

Insectivores, such as hedgehogs and shrews, have to break up the tough chitinous bodies of insects. As this type of diet is not nutritious, insectivores are small creatures. Hedgehogs typify this dentition, where the teeth have lots of sharp, pointed cusps that fit into pits in the opposing teeth and act like a pestle and mortar (Figure 4.22). Curiously, some shrews, such as the red-toothed shrews, have iron pigment in their teeth, giving them a red colour. As the pigment is concentrated at the tips of the teeth (Figure 4.23), some scientists believe that the iron makes the teeth harder, although that would not explain the success of all the white-toothed shrews that lack the iron pigment.

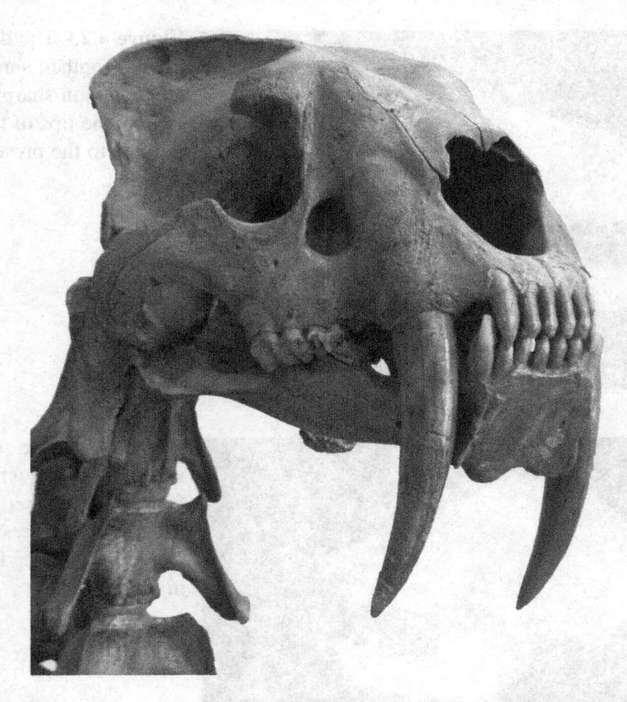

Figure 4.21 The skull of the extinct sabre-toothed cat (*Smilodon*). Note the huge size of the upper canines and the blade-like slicing teeth at the back of the jaw.
Source: http://en.wikipedia.org/wiki/File:Smilodon_head.jpg. This file is licensed under the Creative Commons licence http://creativecommons.org/licenses/by-sa/3.0/deed.en and the GNU Free Documentation license http://en.wikipedia.org/wiki/GNU_Free_Documentation_License

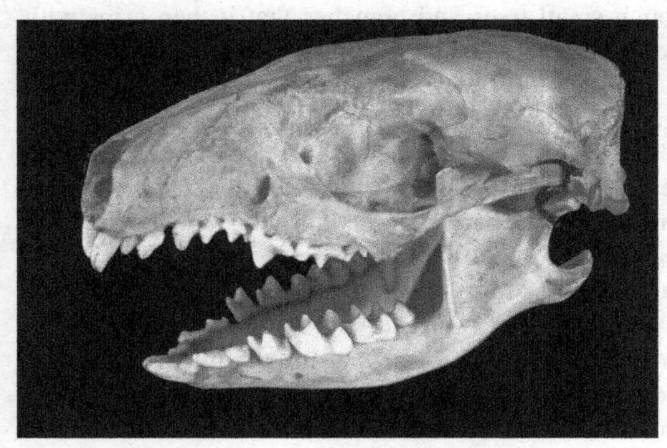

Figure 4.22 Skull of a hedgehog. Note the molar teeth bristling with lots of sharp cusps necessary to crush the hard, chitinous skins of insects.

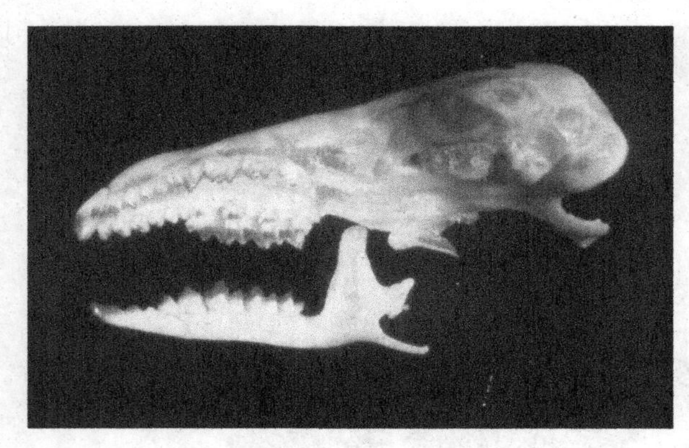

Figure 4.23 The dentition of the red-toothed shrew. The teeth contain sharply pointed cusps. The tips of the teeth are red due to the presence of iron pigment.

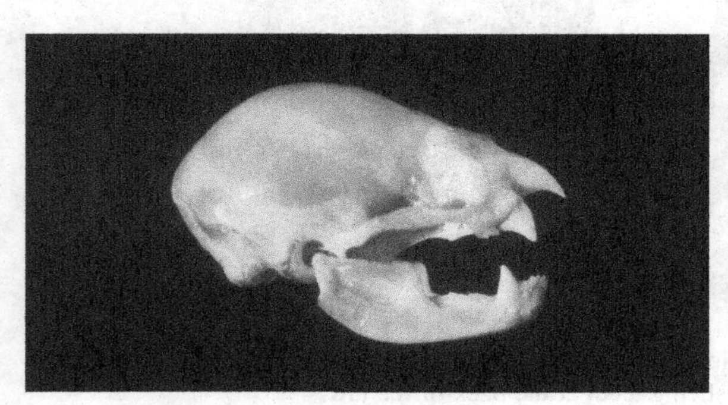

Figure 4.24 The skull of a vampire bat showing enlarged front teeth for piercing the skin of its prey. The rest of the teeth are greatly reduced in size and in number.

Some Other Functions of Teeth

The canine teeth in the males of some species, such as baboons, are enlarged and are used directly as offensive weapons when fighting. They can be used indirectly in threat gestures to frighten off competing males without the necessity of fighting. (The topic of tusks is considered separately in Chapter 1.)

The sharp-edged, continuously growing incisors of the beaver (see Figure 9.18) allow the animal to gnaw through small trees and branches that it then grips with its teeth and uses to construct dams and as a source of food.

In vampire bats, two upper front teeth on each side (representing an incisor and canine) have evolved into large, sharp and pointed teeth that are used to pierce the skin of its prey to promote blood flow, which is then lapped up. The rest of the teeth are small and functionless (Figure 4.24) due to its liquid diet. Like leeches (see pages 58–60), the saliva of vampire bats contains an anticoagulant to slow down the rate of blood clot formation in the host wound.

In the lemurs of Madagascar, the lower four front teeth on each side are grouped together and lie horizontally (instead of vertically) to form a tooth-comb (Figure 4.25A). This tooth-comb is used both in self and in communal grooming, an important social activity bonding these creatures together. That grooming also has an important role to play in the health of the animal has been demonstrated in a

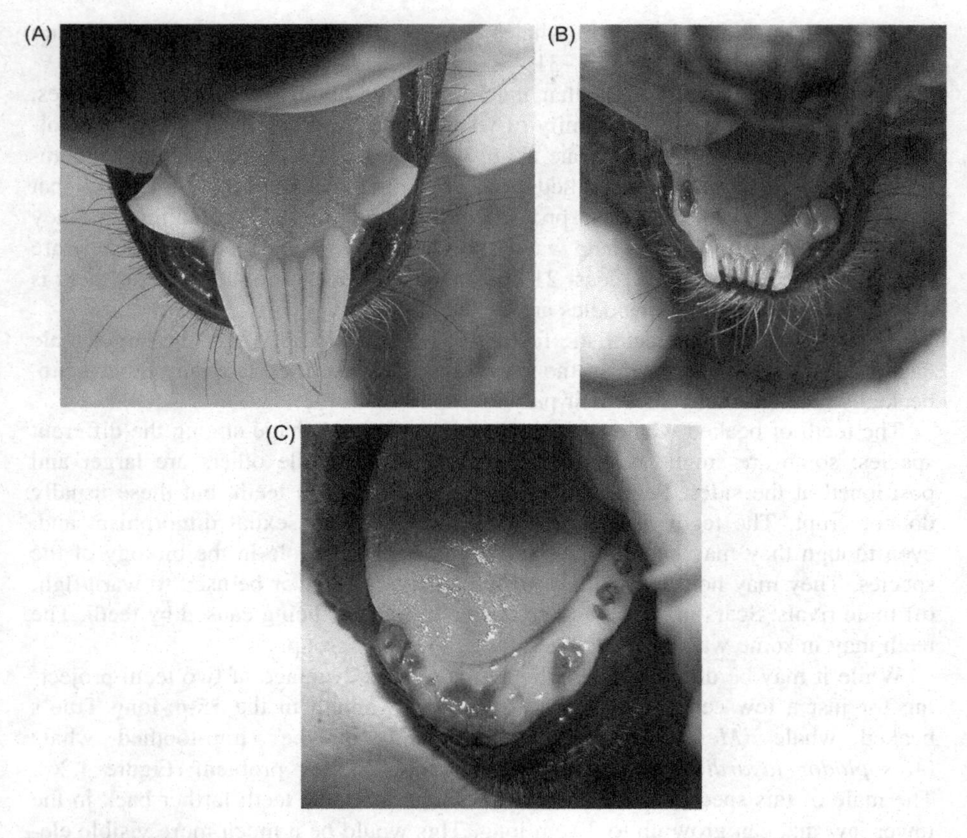

Figure 4.25 Illustrations showing different degrees of wear in the tooth-comb of a lemur. A, unworn tooth-comb; B, moderately worn tooth-comb; C, heavily worn tooth-comb. *Source*: Courtesy of Dr M. Sauther.

recent study, which showed that the tooth-comb wears away or gets damaged in older animals (Figure 4.25B and C). Grooming helps keep down the numbers of parasites, such as mites. When lemurs groom each other, an individual with worn-down or damaged teeth may not be able to groom other lemurs efficiently. This seems to be recognised as in return, such a lemur with worn teeth receives inefficient grooming from other lemurs and eventually ends up with greater infestation of its coat with parasites: the more tooth wear, the more mites.

Whales' teeth differ from the general pattern described for mammals. There is only a single set of teeth present that lasts for life. In dolphins and porpoises (that eat a fish diet, piscivorous), the teeth all have a similar and simple cone shape as the prey is swallowed whole without chewing. Sperm whales have large teeth with a simple cone shape in the lower jaw. These teeth lack enamel and the remaining dentine is considered as ivory (see Chapter 2, pages 31–33). Killer whales have similar teeth, but in both jaws.

The most specialised teeth in whales belong to male narwhals (their tusks are described in Chapter 2, pages 26–31).

Although it may be thought that much is known about the biology of whales, this is far from the truth. The family of whales with the most species after the dolphins and porpoises is also the one about which least is known. This family comprises the beaked whales, so-called because, unlike the majority of whales that have a flattened head, their head projects in a manner similar to dolphins. As they spend much of their time in deep water and are not particularly numerous, they are rarely observed. There are at least 21 known species, but information about them is only ascertained when their bodies are washed ashore.

A specialised and characteristic feature of beaked whales is that the larger male has but a single pair of teeth in the lower jaw. Because they lack any other teeth, beaked whales can only suck their prey into the mouth.

The teeth of beaked whales vary in position, size and shape among the different species: some are small and positioned at the front, while others are larger and positioned at the sides. Females also have a pair of small teeth, but these usually do not erupt. The teeth of the male therefore illustrate sexual dimorphism and, even though they may be small, must play an important role in the biology of the species. They may help individuals recognise one another or be used to warn/fight off male rivals. Scars in the skin have been attributed to being caused by teeth. The teeth may in some way be attractive to females in the group.

While it may be difficult to see the evolutionary advantage of two teeth projecting for just a few centimetres at the front of the mouth in the 15-m-long True's beaked whale (*Mesoplodon mirus*), the teeth of the strap-toothed whale (*Mesoplodon layardii*) are much larger, yet pose another problem (Figure 4.26). The male of this species develops a pair of large tusk-like teeth farther back in the lower jaw that can grow up to 30 cm long. This would be a much more visible element to use for signalling, fighting and attracting females. The problem is that, because of their inward slope, the teeth grow to surround the upper jaw and eventually physically restrict the whale from opening its mouth more than about 10 cm.

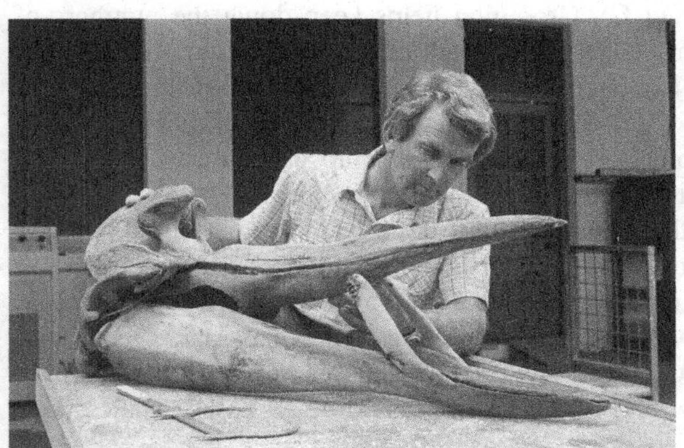

Figure 4.26 Skull of a strap-toothed beak whale. Note the pair of curved tusks growing upwards from the lower jaw. *Source*: Courtesy of the Alexander Turnbull Museum.

This whale's diet becomes reduced to eating very small fish and squid, and yet this condition is part of its normal biology. Amazing!

Functionless Teeth

Although this chapter has taken for granted that all teeth erupt into the mouth and have a function, in some situations, teeth may initially develop within the jaws but never erupt and may be lost without ever being used. An extreme example of reduced teeth having no function is seen in alligators. During their lifetime, alligators may have as many as 30 sets of replacing teeth. However, before they have even hatched from the egg, four sets of tiny teeth (each set slightly larger than the previous one) have already developed and been replaced while it is still inside the egg. Thus, the set of teeth the animal has ready to use when first hatched actually represents its fifth set.

The dentitions of pouched marsupials in Australia are very different when compared with the rest of the mammals in that they have only one set of teeth, apart from a single replacing tooth in the middle of the row. However, many marsupials, such as wallabies and tree-dwelling phalangers, have a number of tiny teeth which develop very early on but rapidly disappear without ever functioning.

The guinea pig, a familiar cuddly pet, possesses a molar tooth in each jaw that develops early but is lost before birth. What is unusual in this case is that the biting surface of the tooth shows evidence of tooth wear, indicating that the guinea pig carries out tooth-grinding while still in the womb.

Whales provide examples of functionless teeth. The male narwhal has, in addition to its main tusk (usually the left tooth; see Figure 2.22), a very reduced equivalent tooth on the right side of the skull that remains unerupted. Female narwhals have two small, vestigial tusks that remain unerupted throughout life. The sperm whale possesses a single set of 40–50 large, conical teeth in the lower jaw, but in the upper jaw, a number of small teeth develop but never erupt (Figure 4.27).

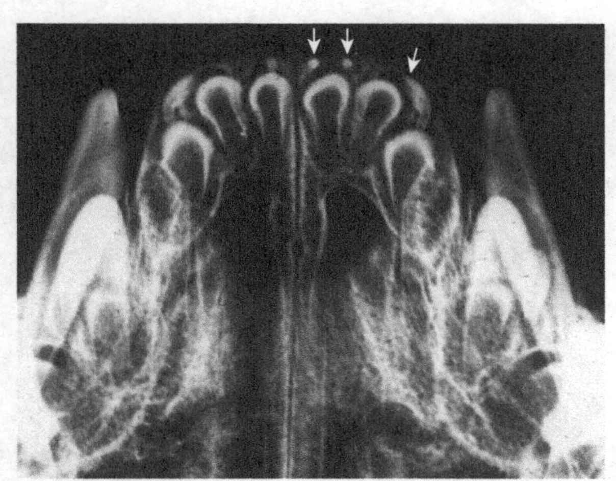

Figure 4.27 X-ray of a young ferret (*Putorius putorius*). The functionless deciduous incisors are arrowed. The larger permanent incisors are seen developing behind them.

The milk front teeth of small carnivores such as the cat and ferret are tiny and vestigial. An X-ray will reveal their presence beneath the gum at birth, but they are soon lost without ever functioning (see Figure 9.27).

These examples of vestigial teeth are presumably the last remnants of teeth once functional during an earlier stage in the evolutionary history of the animal.

5 Teeth in the Most Unlikely Places

Unusual Medical Case History 1

A young woman aged 30 visited her doctor complaining of slight, intermittent pain on the right side of her abdomen. The doctor felt a small enlargement at the site of the pain and requested that an X-ray be taken. The doctor later showed the X-ray to the patient (Figure 5.1) and said 'I have good news and bad news. The bad news is that the X-ray confirms my clinical diagnosis, namely that you have a tumour in your right ovary. This is evident from a round, dark area surrounding the central mass and suggesting it is filled with fluid. The good news is that, even without a biopsy, I am confident the tumour is benign. How can I be sure of this? Well, you can see that within the central part of the tumour there are about twelve small white structures. On an X-ray, anything white is hard, such as a bone. In this case, because of their shape, the white structures in your tumour are teeth'.

Subsequently, the tumour (also known as an ovarian teratoma or dermoid cyst) was removed and found to contain a number of teeth.

Teeth in the Tummy

Accounting for between 25% and 50% of all tumours in the ovary, teratomas are the most common type. In normal fertilisation, a sperm penetrates an egg (ovum) released from the ovary. This egg then becomes attached to the wall of the uterus where development of the baby begins. In the case of an ovarian teratoma, the egg starts to divide and grow uncontrollably within the ovary, without being fertilised. The mechanism of this process, known as parthenogenesis, remains unknown. As the egg cells have the full complement of genes and the potential to make a complete individual, it is not surprising that many different tissues may be found within the tumour, although haphazardly arranged. Many of the tissues are skin-derived, such as hair follicles, sebaceous and sweat glands, nails and frequently teeth. In addition, cartilage, bone, thyroid gland and even brain tissue may also be found. The tumour can vary in size, from small and asymptomatic to a very large mass causing swelling and pelvic pain. The condition is most common in women aged between 20 and 30 years. Smaller, less common, dermoid cysts may also occur in male testes.

Although generally occurring in one ovary, an ovarian teratoma affects both ovaries in about 10% of cases. The presence of readily identifiable tissues in well over 90% of ovarian teratomas means that the tumour is nearly always benign.

Nothing but the Tooth. DOI: http://dx.doi.org/10.1016/B978-0-12-397190-6.00005-5

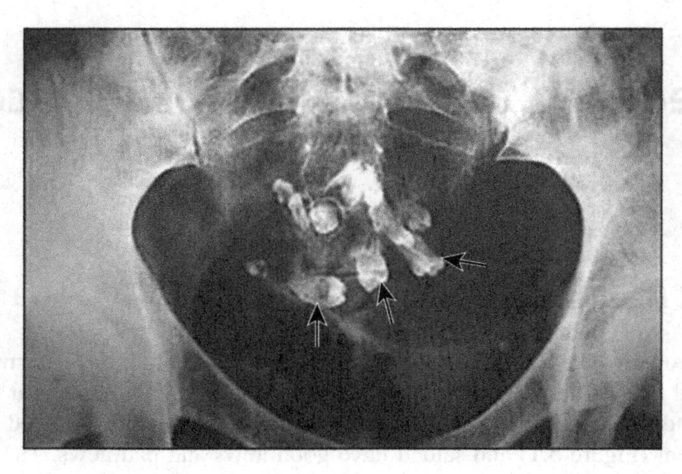

Figure 5.1 X-ray showing an ovarian cyst within which can clearly be seen at least 12 teeth (some indicated by arrows), whose white appearance confirms they are highly mineralised. *Source*: Courtesy of Dr C.M. Peterson at http://library.med.utah.edu/kw/human_reprod/mml/hrovt_xray01.html.

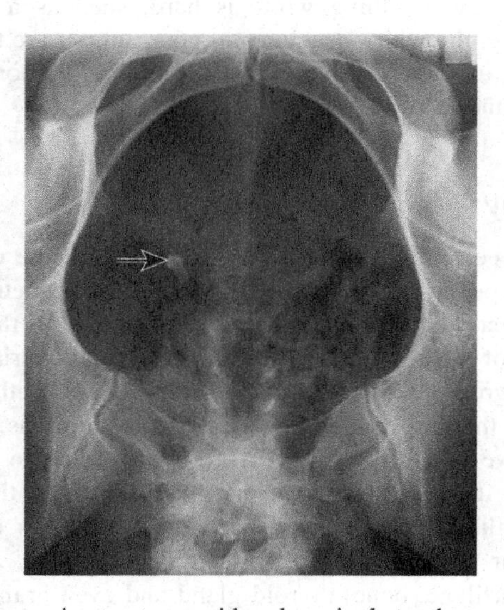

Figure 5.2 X-ray of an ovarian teratoma with only a single tooth present (arrow). *Source*: Courtesy of Dr J. Luker.

Rarer (10%), malignant tumours contain more 'primitive' (undifferentiated) cells, which would not have been able to form such a complex organ as a tooth.

When teeth are present in ovarian teratomas, the number may vary considerably. Sometimes, only one or two are present (Figure 5.2). In other cases, several may be

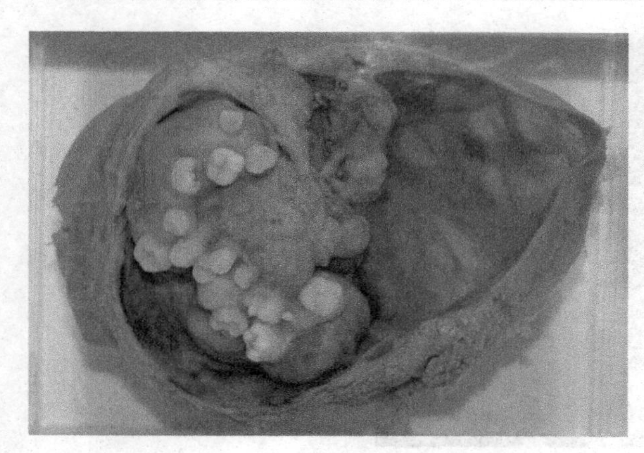

Figure 5.3 Large ovarian cyst with 18 teeth present.
Source: Courtesy of the Gordon Museum, King's College London.

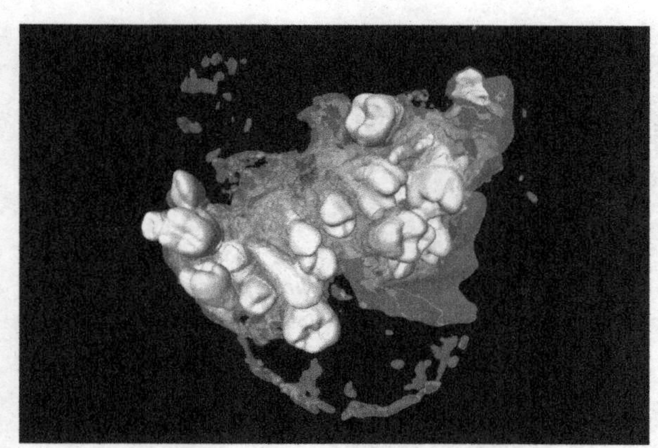

Figure 5.4 Three-dimensional reconstruction following a CAT scan, showing the detailed shape of the teeth of specimen seen in Figure 5.3.
Source: Courtesy of Dr K. Kupczick and the Gordon Museum, King's College, London.

found. The very large teratoma shown in Figures 5.3 and 5.4 had a diameter of 10 cm and possessed 18 well-formed teeth. The roots of some teeth may even be surrounded by a bony socket. Very occasionally, the teeth are found sitting in a rudimentary jaw. Figures 5.5 and 5.6 show such a case where the teratoma was large (measuring over 15 cm in diameter and weighing 1.8 kg) and possessed a rudimentary jaw containing 14 teeth.

The teeth that develop in ovarian tumours still show all the tissues (enamel, dentine, cement and dental pulp) and microscopic features seen in normal teeth, even including the presence of growth lines. The presence of these growth lines indicates that, even in the unusual environment of the ovary, tooth development is still under the influence of the daily and weekly body rhythms as in normal teeth (see Chapter 13).

Some teeth have recognisable shapes of real teeth; others have a simple peg shape dissimilar to that of normal teeth, while others may have an intermediate shape. The majority of teeth in ovarian tumours resemble adult premolars and molars, and a minority resemble incisors and canines. Very few resemble milk teeth (see Chapter 9). The reasons for this distribution of shape are not known.

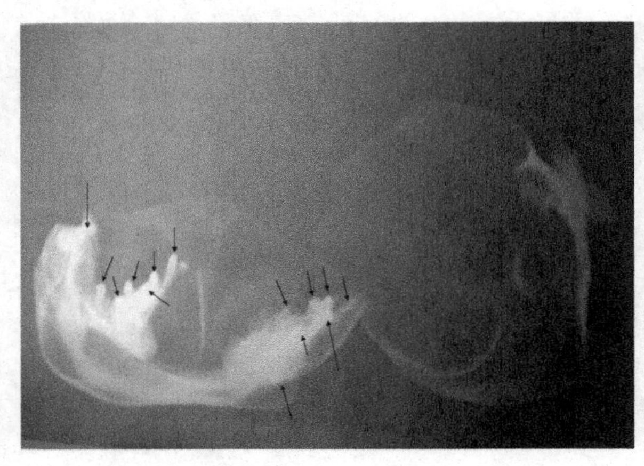

Figure 5.5 X-ray of an ovarian teratoma in which a rudimentary jaw has developed with 14 teeth present within it. *Source*: From Drs S.S. Chavan and V.V. Yenni, 2009. Courtesy of the editors of the *Indian Journal of Pathology and Microbiology*.

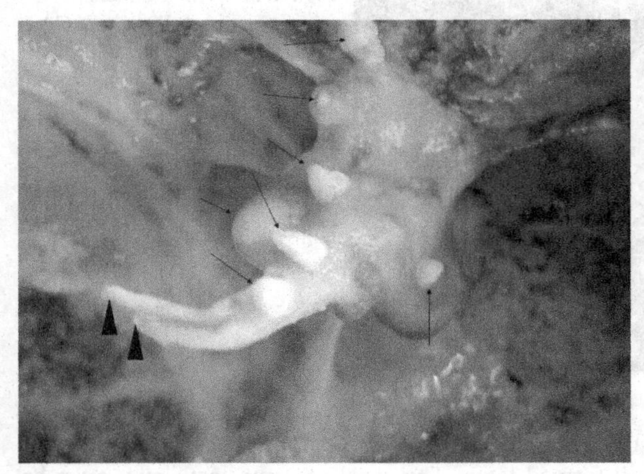

Figure 5.6 Part of the rudimentary jaw dissected from the ovarian teratoma seen in Figure 5.5, showing seven teeth (small arrows) and two horn-like projections (large arrow heads). *Source*: From Drs S.S. Chavan and V.V. Yenni, 2009. Courtesy of the editors of *Indian Journal of Pathology and Microbiology*.

In normal teeth, the growth rate of enamel differs between milk and permanent teeth, with the milk teeth forming at a faster rate. In ovarian teratomas, the growth rates of the teeth are slow, being similar to those of permanent teeth, even in those teeth whose shape more closely resembles milk teeth. This might indicate that the teeth in teratomas start to develop at the same time as the normal permanent teeth (i.e. after birth) rather than before birth, as is the case for milk teeth.

Unusual Medical Case History 2

During a teaching session in the ophthalmology department of a hospital, a student was asked by the consultant to look carefully at the left eye of two patients and write a report of the findings.

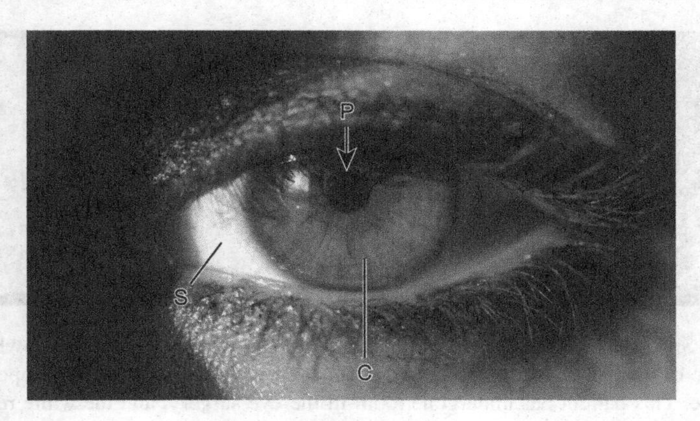

Figure 5.7 Appearance of a normal eye. Note the white of the eye (sclera, S), the transparent cornea (C) covering the coloured iris and the pupil (P).
Source: http://en.wikipedia.org/wiki/File:A_woman's_eye.JPG. This file is licensed under the Creative Commons license http://creativecommons.org/licenses/by-sa/3.0/deed.en and the GNU Free Documentation License http://en.wikipedia.org/wiki/GNU_Free_Documentation_License

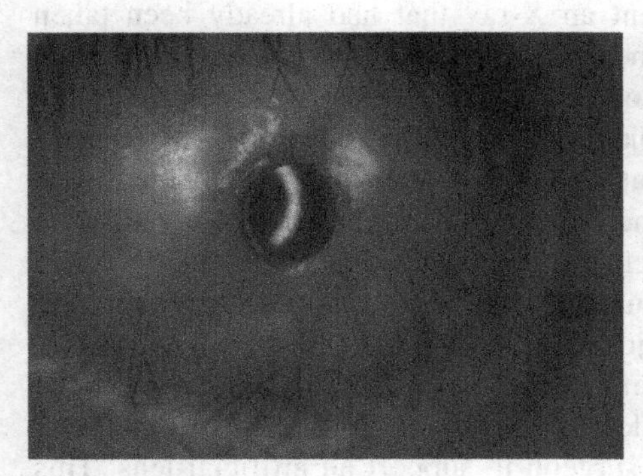

Figure 5.8 Abnormal appearance of the eye. Note the differences compared with normal eye in Figure 5.7. This eye is pink in colour and has no white area or an iris or pupil. This patient has previously undergone tooth-in-the-eye surgery.
Source: Courtesy of Professor C.S.C. Liu.

The student examined the left eye of the first patient, which seemed normal. The outer part was white, while the central part contained a brown iris. At the centre of the iris was the black circle of the pupil (Figure 5.7) whose margins characteristically constricted when light was shone into it.

Turning to the left eye of the second patient, the student was disconcerted by its unexpected appearance. Instead of being white, the eye was slightly pinkish throughout. There was no sign of a pigmented iris surrounding the pupil. Instead, the central region of the eye was occupied by a cylindrical structure with raised edges that projected slightly out of the eye (Figure 5.8). This strange 'pupil' did not constrict in response to light, giving the patient a robotic, staring look. Asked by the consultant to comment, the student was at a loss for words, never having

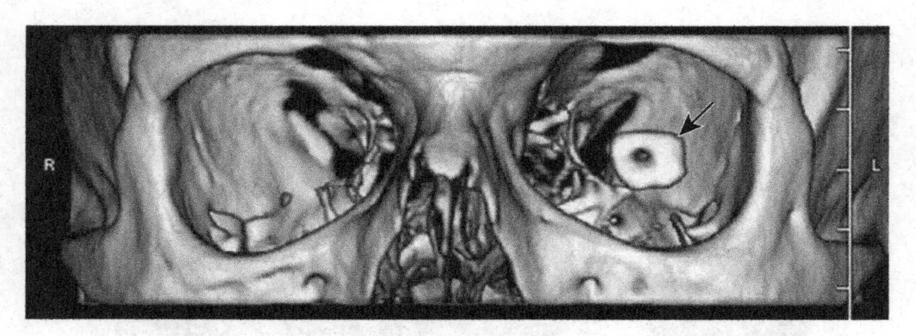

Figure 5.9 Special type of X-ray (electron beam tomography) showing a dense white region in the centre of the patient's left eye (arrow) compared with the normal, clear appearance of the right eye. This patient has undergone tooth-in-the-eye surgery, and the white region represents the tooth root/ocular cylinder.
Source: Courtesy of Professor C.S.C. Liu.

seen anything remotely like it, and could only suggest that perhaps some object had become lodged in the eye following an accident. When asked by the consultant what clinical test should be carried out next, the student suggested taking an X-ray of the eye.

The consultant then passed the student an X-ray that had already been taken (Figure 5.9) and asked for further comment. The student pointed out that, unlike the patient's normal right eye, the affected left eye revealed a white object in the centre. On being asked what the white object might be, the student correctly replied that the white appearance indicated that the object was composed of a dense material, such as a radiopaque dye or a metal. Asked what human tissue could give a similar appearance, the student correctly suggested bone. The consultant then asked the student to name another human tissue that had a similar appearance to bone, but was even denser. The student, jokingly, referred to the teeth. The consultant was pleased with this answer and informed the student that the object in the centre of the patient's right eye was indeed part of a tooth and that it had been deliberately implanted into the eye by a surgeon to support an artificial lens. This specialised surgical procedure had miraculously restored the patient's sight after a period of 20 years of blindness in that eye, there having been no previous cure. The procedure is known as an osteo-odonto-keratoprosthesis (from the Latin osteo = bone, odonto = tooth, kerato = cornea, and prosthesis = artificial device) or OOKP for short. It is known more simply as 'tooth-in-the-eye' surgery.

How We See

There are three basic layers to the eye Figure 5.10: the outermost, white layer (the sclera); the middle, vascular layer (the choroid) and the innermost, nerve layer (the retina). In normal vision, light passes through the most superficial layer at the centre of the eye, the cornea, which is transparent and lacks blood vessels. At its

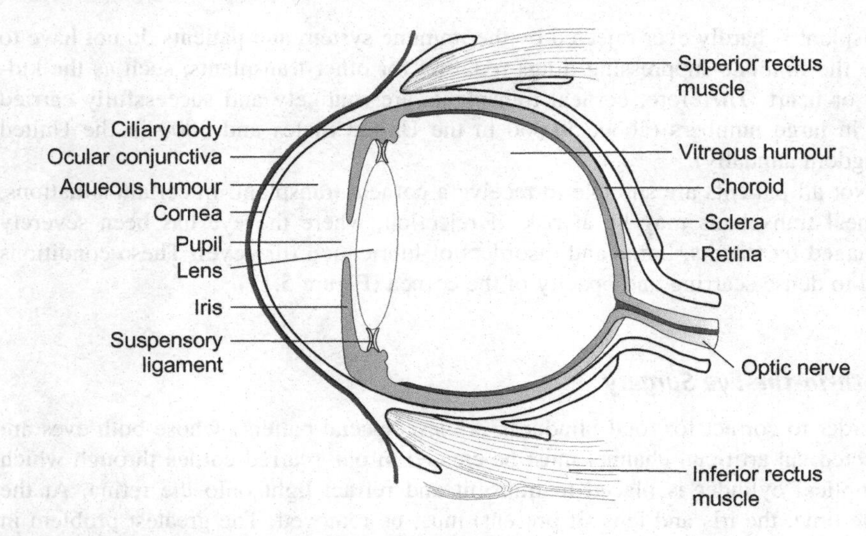

Figure 5.10 Diagram of visual pathway in normal eye. See text.

periphery, the cornea is continuous with the white part of the eye (the sclera). The light continues through the pupil which is an opening in the coloured part, the iris. The iris itself is the front part of the middle layer of the eye, the choroid.

The amount of light passing through the pupil is controlled by muscles in the iris, which cause it to constrict or dilate like the lens of a camera. Light then passes through the lens, where it is bent to fall directly onto the innermost and most sensitive part of the retina at the back of the eye. The lens is supported by a suspensory ligament, and it can change its curvature following contraction and relaxation of ciliary muscles in the ciliary body. This means the eye can accommodate to either close images (while reading) or distant objects.

The light rays are focused on the light-sensitive cells of the retina to produce nerve impulses that pass to the brain, where they are processed to give us the property of vision. The space in front of the lens is occupied by fluid (aqueous humour) while that behind the lens is filled with a jelly-like substance (vitreous humour). The function of these fluid/jelly-filled spaces is to support and protect structures in the eye.

Corneal Blindness

One of the most common causes of blindness is the loss (partial or complete) of the normal transparency of the cornea, which becomes opaque and prevents light from passing through the lens and reaching the retina. Blindness may also result from a distorted cornea. As long as the retina is normal, these types of blindness can be cured by means of a corneal transplant. In this procedure, the cornea of a recently deceased donor is exchanged for the damaged cornea of the recipient. This type of transplant is hardly ever rejected by the immune system and patients do not have to take the immune-suppressing drugs required for other transplants, such as the kidney or heart. Therefore, corneal transplants are routinely and successfully carried out in large numbers (about 50,000 in the United States and 3000 in the United Kingdom annually).

Not all patients are suitable to receive a corneal transplant. In certain situations, corneal transplants may be at risk of rejection, where the eye has been severely damaged by trauma, burns and disorders of lubrication (dry eye). These conditions lead to dense scarring and opacity of the cornea (Figure 5.11).

Tooth-in-the-Eye Surgery

In order to correct for total blindness in these special patients whose both eyes are affected, an artificial channel must be created in one scarred cornea through which an optical cylinder is placed to transmit and refract light onto the retina. At the same time, the iris and lens (if present) must be removed. The greatest problem in these rare and difficult cases that cannot receive a corneal transplant is how to support the optic cylinder within the eye. The most satisfactory solution to date, and the one that has been used successfully for 50 years, is to support the optic cylinder using the root of a tooth and insert the whole device into the affected eye. This remarkable procedure was initially developed in Italy by Professor B. Strampelli and has since been modified and improved by a number of other ophthalmic surgeons. Only a limited number of specialist centres around the world are capable of undertaking this complex operation. In the United Kingdom, one such centre exists in the Sussex Eye Hospital. The procedure was performed in the United States, normally a leader in technological medicine, for the first time in 2010.

Tooth-in-the-eye surgery involves a number of separate procedures undertaken over several months. Only the basic principles are outlined here.

Because the thin, protective layer of the cornea (the conjunctiva) is missing, and there is invariably a deficiency in the lubricating system of tears, the first task is to create a new protective coat for the damaged cornea that will provide lubrication. This is achieved by covering it with a thin layer of moist skin taken from the lining

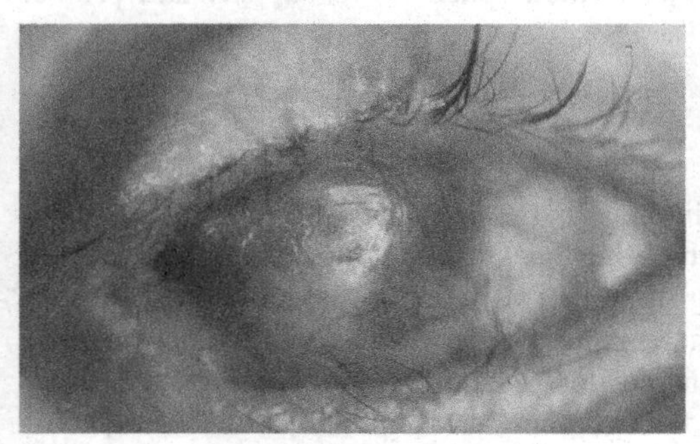

Figure 5.11 An example of corneal blindness due to extreme dryness of the eye and not amenable to conventional corneal transplant treatment.
Source: Courtesy of Professor C.S.C. Liu.

of the mouth, which suitably contains mucous glands and lacks hair. A section of this mouth lining is removed, trimmed to fit over the damaged cornea and sclera (the white of the eye) and sutured over the front of the eye, giving the whole surface a pink appearance (as opposed to the normal white) (Figure 5.12).

While one team of surgeons is involved in operating on the eye, a second team operates on the jaw. A healthy, single-rooted tooth is identified in the patient (preferably the upper canine). This is removed together with its surrounding bone and the intervening (periodontal) ligament (Figure 5.13). The root of the extracted tooth is then cut in half along its length, the central soft pulp removed, and the tooth carefully trimmed to leave an inner surface of dentine and an outer surface of bone (with the ligament remaining in between). A round hole (of the correct diameter in which to fit a clear optic cylinder) is drilled in the middle of the root fragment. This procedure eventually provides a thin (3 mm) rectangular piece of root about 10 mm wide, with a hole in the centre (Figures 5.14 and 5.15). The optical plastic cylinder is then firmly cemented into the hole, creating a 'tooth-lens' (Figure 5.16). To provide a blood supply and keep it vital, the 'tooth-lens' is implanted beneath the skin in a safe region, such as the lower eyelid on the unaffected side.

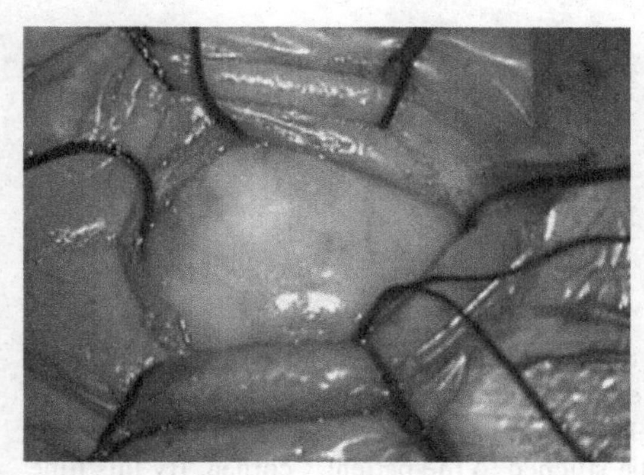

Figure 5.12 A damaged cornea is covered with a graft taken from the inside of the cheek. It appears pink, and there is no evidence of the normal white layer of the eye. The cheek region has the advantages of containing mucous glands for lubrication and lacking hairs.
Source: From A. Gomaa, O. Comyn and C. Liu, 2010. Courtesy of the editors of *Clinical and Experimental Ophthalmology*.

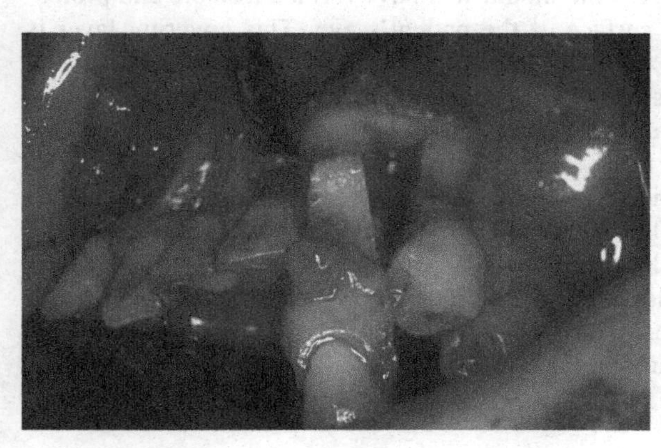

Figure 5.13 Operation to remove a block of the jaw containing the complete lower canine tooth and its surrounding bone.
Source: Courtesy of Professor C.S.C. Liu.

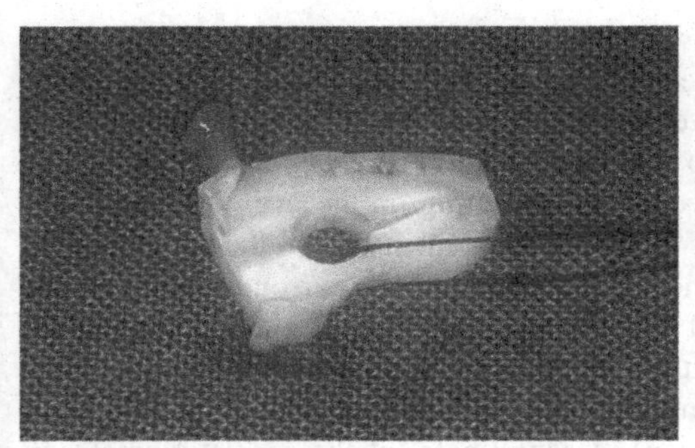

Figure 5.14 Block of the root still attached to the bone by a thin fibrous joint, the periodontal ligament. A hole has been drilled into the root to accommodate an optical cylinder. *Source*: Courtesy of Professor C.S.C. Liu.

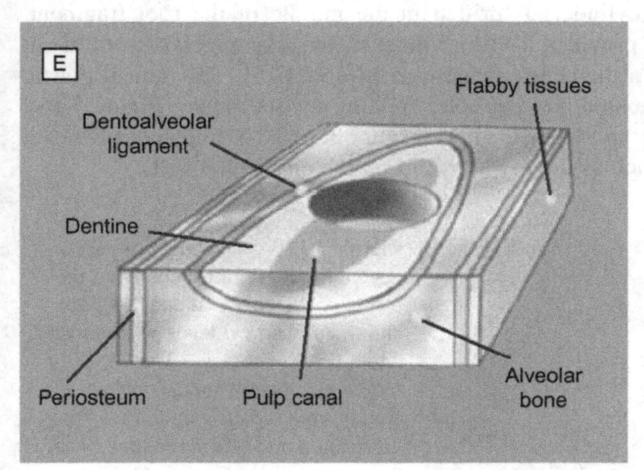

Figure 5.15 Diagram showing labelled root tissues seen in Figure 5.14. *Source*: Courtesy of Professor C.S.C. Liu.

A few weeks later, the patient undergoes a second operation. The implanted tooth-lens has by this time become surrounded by healing, vascular tissue. It is removed, cleaned on its upper and lower surfaces, but the tissues at the edges are retained to allow the tooth-lens to be stitched to the patient's cornea. By this time also, the mucosa transposed from the mouth will have formed a stable and protective layer covering the front surface of the patient's eye. This covering layer is then folded downwards to expose the patient's original cornea beneath, in the centre of which an opening is created big enough to accommodate the tooth-lens. Additional temporary cuts are made at the margins of the pupil to allow the iris and original lens (if present) to be removed. The tooth-lens is then positioned so that its surface sticks out a little from the front of the eye. It is then stitched to the cornea and the mucosal flap is pulled back over it. A small hole is cut in the mucosal flap through which the tooth-lens can protrude (so as not to be overgrown by the protective layer of mouth mucosa) (Figures 5.8 and 5.17). Once the implant has proved successful, the appearance of the tooth-lens can be cosmetically improved by adding a false iris.

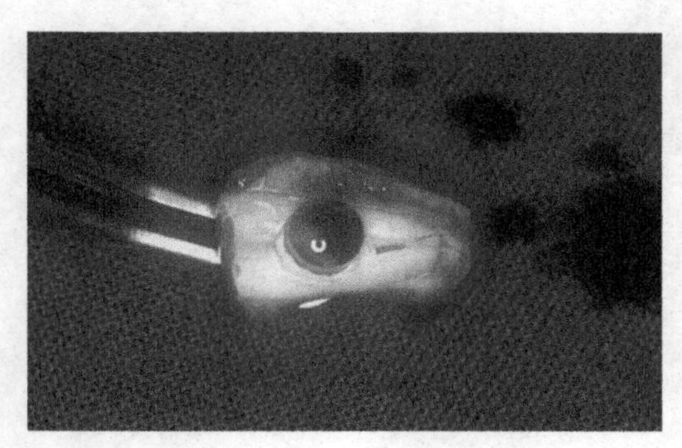

Figure 5.16 A plastic optical cylinder is cemented into the hole in the root.
Source: Courtesy of Professor C.S.C. Liu.

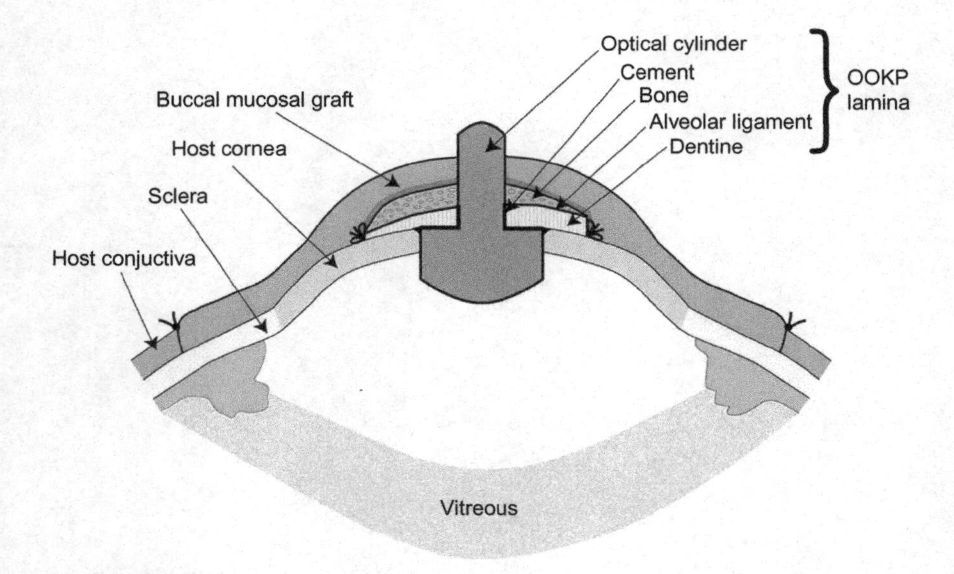

Figure 5.17 Diagram showing final position of the prosthetic appliance following tooth-in-the-eye surgery.
Source: From A. Gomaa, O. Comyn and C. Liu, 2010. Courtesy of the editors of *Clinical and Experimental Ophthalmology*.

If the patient has no teeth, it is possible to use the tooth of a close relative with the same matching blood group, although this is less likely to be successful due to a greater chance of transplant rejection.

Research is continuing to find a better alternative than the patient's own tooth to support the optical cylinder. Although a few biological materials have been experimented with, such as coral, none have yet been adopted to replace the tried and tested root of a tooth.

6 Snakes, Saliva and a Nobel Prize

Scientists work in virtual anonymity. The results of their research are presented at conferences whose attendees may number from just a handful to upwards of a thousand or more delegates. The respect and admiration of their peers are generally sufficient to motivate them. Recognition comes with the prestige of being the first to make a major discovery. There is little glory in being second! The only scientists known to the general public are the handful that appear on television or who write a best-selling book on popular topics such as astronomy, evolution or genes. However, every year one or two scientists are plucked from the obscurity of academia and brought to the attention of the general public. That moment comes when they are awarded a Nobel Prize. From then on they attain superstar status, and their lives are changed forever.

The award is named after Alfred Nobel, a Swedish scientist whose fortune resulted from his discovery of dynamite, which he adapted for use in controlled explosions. Although expecting dynamite to be used for peaceful engineering projects, he was deeply disturbed to find it being used in war. To compensate for his feeling of guilt, Nobel decreed that his fortune be used to establish annual prizes recognising major achievements that benefited mankind. The five original categories of Nobel prizes, first awarded in 1901, were in the fields of Medicine or Physiology (awarded by the Karolinska Institute), Physics and Chemistry (both awarded by the Royal Swedish Academy of Sciences), Literature (awarded by the Swedish Academy) and Peace (awarded by the Norwegian Parliament). A sixth Nobel Prize in Economics was created in 1969 (awarded by Sveriges Riksbank, Sweden's central bank).

The Nobel Prize is considered the most prestigious in the world. To reflect this, the winners receive a diploma and gold medal from the ruling monarch of Sweden in an impressive ceremony in Stockholm, together with a cash award whose present day value is in the region of £750,000 ($1,200,000). The prize is often awarded many years after the initial discovery to ensure that its importance has been validated. For example, a discovery as immediately significant as the double helix of DNA in 1953 was not recognised for a Nobel Prize until 1962, the recipients being Drs F.H.C. Crick, J.D. Watson and M.H.F. Wilkins. Unusually, the 1923 award for insulin to Drs F.G. Banting and J. Macleod was made only 1 year after its discovery.

As the Nobel Prize is not awarded posthumously, it behoves potential recipients to be prepared to live a long life, as in the case of Dr P. Rous. He was awarded the 1966 Nobel Prize for Medicine at the age of 87 for a discovery he made in 1910,

Nothing but the Tooth. DOI: http://dx.doi.org/10.1016/B978-0-12-397190-6.00006-7

nearly 60 years earlier, concerning the involvement of viruses in the transmission of certain types of cancer. Conversely, Dr Rosalind E. Franklin, who made a major contribution towards the discovery of the structure of DNA, died in 1957 at the age of 37, 4 years before the Nobel Prize was awarded.

At the Nobel Prize ceremony, each recipient gives a public lecture in which they describe the work that led to their award and refer to scientific papers that demonstrated their important findings. For one Nobel Prize (in Medicine), teeth played a small but important part in the discovery, and it is the story of this prize that is recounted here. This discovery revolutionised our understanding of how cells function. Prior to this, little was known about how cell growth and development were controlled or why, for example, some cells give rise to bone while others give rise to nerves or muscles. It was known that endocrine hormones, such as growth hormone from the pituitary gland and thyroxine from the thyroid gland, are released into the blood stream and travel around the body to affect cell behaviour at more distant sites: local control mechanisms were barely understood. This Nobel Prize involved the work of three main people, although only two were to share it.

The first person is Dr Rita Levi-Montalcini, born in Italy in 1909 (Figure 6.1). Throughout her life, she has displayed enormous courage and a dogged determination to fulfil her destiny. Although a little late in choosing medicine as a career, she put in the extra study that allowed her to qualify at the University of Turin in 1936. Remarkably, among the small cohort of students with her at the time, two others were also destined to become future Nobel Prize winners in Medicine, namely Drs S.E. Luria (1969) and R. Dubecco (1975).

As part of her studies, Levi-Montalcini began a research project into the development of the nervous system in chick embryos. Unfortunately, this work was soon curtailed by the fascist government of Benito Mussolini, who decreed that, being Jewish, Levi-Montalcini could not be allowed to pursue her academic career. Things got much worse shortly afterwards when Italy became an ally of Germany during World War II and German troops entered Italy. Levi-Montalcini and her family were forced to live in hiding throughout the war. Despite these severe restrictions, she continued her research as best she could. She constructed a

Figure 6.1 Dr R. Levi-Montalcini. Taken about 1963.
Source: Courtesy of Becker Medical Library, Washington University School of Medicine.

primitive laboratory in her bedroom with very basic equipment and even managed to publish two scientific papers during this period. These appeared in a Belgian scientific journal, as Jews were not allowed to publish in Italian scientific journals.

In 1946, Levi-Montalcini received an invitation to visit St Louis, Missouri, and work with the second figure in this story, Dr Viktor Hamburger (Figure 6.2). He was aware of Levi-Montalcini's research interests, which coincided with some of his own. Hamburger was an eminent researcher who had already made a number of important findings concerning factors influencing the developing nervous system. Being Jewish, he too had to flee from Nazi Germany in 1933 after being dismissed from his university post.

In 1948, a member of Hamburger's team, Dr E. Bueker, published a paper describing observations he made when pieces of a mouse tumour were grafted into the developing hind limb of a chick embryo (while still in the egg). He observed that, compared with the normal side, there was enlargement of the nerves on the tumour side. He concluded that the rapidly growing tumour cells stimulated the growth of nerves in the chick.

Hamburger and Levi-Montalcini confirmed and extended these original findings in new experiments. They observed that not only did nerves grow from the chick embryo into the mouse tumour but also that more distant to the tumour, the chick's own nerves proliferated wildly. Thus, structures such as the chick's veins, which normally possessed few nerves, became virtually blocked by nerves. Their results indicated that the mouse tumour cells did indeed release a 'nerve growth-promoting factor' that produced abnormal changes in the nerves of the developing chick.

This conclusion gained further support in experiments undertaken by Levi-Montalcini when she teamed up with a colleague who had expertise in tissue culture. In this technique, cells can be grown in a test tube when provided with suitable nutrients. In these studies, Levi-Montalcini allowed nerves taken from chick embryos to develop close to mouse tumour cells. This resulted in a dramatic increase in the growth of nerves compared with cultures that did not receive tumour cells. Also, she noticed that the closer the nerves and tumour cells, the stronger the

Figure 6.2 Dr V. Hamburger.
Source: Courtesy of Becker Medical Library, Washington University School of Medicine.

reaction. A paper published in 1954 describing this work provided strong support to the view that the mouse tumour cells released a diffusible factor that stimulated the growth of nerves.

The third person to join the team in 1951 was Dr Stanley Cohen (Figure 6.3), whose parents were Jewish refugees from Russia. As a biochemist, it was his task to isolate and purify the very small amount of active chemical ingredient from the mouse tumour that stimulated nerve growth. By 1954, Cohen, Levi-Montalcini and Hamburger were able to report some success. The active ingredient isolated from the tumour, later to be called nerve growth factor (NGF), had the characteristics of a nuclear protein, i.e. a protein linked to nucleic acid.

The next problem was to determine which of the two components, either the nucleic acid or the protein, was responsible for stimulating nerve growth. This required finding a way to split the two components. During a discussion with a colleague who was an expert on nucleic acids, Cohen learnt that snake venom contained (among many other substances) an enzyme (phosphodiesterase) that broke down and inactivated nucleic acid. He therefore decided to add small quantities of snake venom to the culture medium. If the nerve growth-promoting effect was still apparent, then it must be due to the protein, as the nucleic acid would have been degraded by the snake venom. Conversely, if there was no stimulation of nerves following the addition of the snake venom, the nerve growth activity must be due to the nucleic acid.

The results of the experiment involving the addition of snake venom proved to be startling and totally unexpected. The first positive result was that the addition of snake venom did not diminish the nerve growth-inducing effects of the mouse tumour extract. This indicated that the active ingredient was the protein component of the nucleoprotein extract (and not the nucleic acid). Even more remarkable was the observation that the addition of snake venom alone increased nerve growth by a thousand times. Totally by chance (serendipity), the research team had stumbled upon the fact that snake venom itself contained large quantities of NGF and could be more conveniently used as a source of NGF rather than mouse tumours.

While using snake venom, Cohen searched for a more readily obtainable and cheaper source of NGF. The poison gland in the snake is equivalent to the major salivary glands in mammals. Cohen soon found a rich source of NGF in one of the salivary glands of male mice, the submaxillary gland (although, somewhat surprisingly, little is present in the female). From this source he eventually extracted

Figure 6.3 Dr S. Cohen. Taken in 1986.
Source: Photo by Mr D. Wile.

enough NGF to characterise its chemical composition and to inject it into live animals.

The unusual and seemingly haphazard distribution of NGF, being found in mouse tumours, snake venom and male mouse salivary glands, did not immediately seem to implicate it as being an important protein. However, when researchers blocked its activity in developing embryos, major disturbances were produced in the nervous system. This provided evidence that NGF was important during the early development of the nervous system.

The final twist in the story came when Cohen had eventually purified enough of the crude extract of the protein to inject into live, newborn mice. As predicted, he observed an unusually large excess of nerve growth activity, finally proving that NGF was an important and fundamental molecule controlling development of the nervous system. He was also observant enough to notice two relatively minor features which had nothing to do with nerves (another case of serendipity) in newborn mice receiving the protein extract: (i) the eyes opened earlier (at 6–7 days after birth compared to the usual 12–14 days) and (ii) the incisor teeth erupted earlier (at 5–6 days after birth compared to the usual 8–10 days). These two differences may seem small and relatively unimportant. While most scientists probably would have disregarded or not even noticed them, Cohen pursued them further. To make sure they were not simply chance findings, he repeated the experiment and confirmed they were highly reproducible. In a flash of inspiration, he interpreted the results as indicating that, in addition to NGF, the male mouse salivary gland contained an additional and entirely different and new growth factor. Due to its effect on speeding-up growth and development of the skin (epidermis) of the eyelid, he named it epidermal growth factor (EGF). In 1962, having isolated this second growth factor, he wrote a paper describing his results. The full title of this paper, which appeared in Volume 237 of the *Journal of Biological Chemistry* (pages 1555–1562), was 'Isolation of a Mouse Submaxillary Gland Protein Accelerating Incisor Eruption and Eyelid Opening in the New-born Animal'. Thus, the early eruption of the incisor teeth was a key observation in the discovery of EGF.

Subsequently, Levi-Montalcini, working with NGF, and Cohen, working with EGF, provided much data indicating the fundamental importance of these two growth factors during cell development and behaviour. It was shown that they are produced by a variety of cells and interact with receptors on the surface of cells. This interaction leads to a cascade of complex biochemical reactions inside the cell, controlling its growth, development and behaviour. The discovery of the first two growth factors opened the floodgates. Since that time, a large number of other growth factors have been identified. If growth factors released by white blood cells during inflammation are also included, the number reaches into the hundreds. Today, an Internet search keying in the words 'growth factors' would produce over 50,000,000 hits.

Since their discovery, growth factors are known to play a pivotal role in the normal growth and development of all tissues in the body. For example, NGF has important functions in the adult nervous system associated with learning and memory. Any imbalance in the production of growth factors is likely to be associated with disease. Knowledge concerning growth factors is fundamental to understanding and treating a wide range of diseases, from Alzheimer's to cancer.

In recognition of their discovery of the first two growth factors, Levi-Montalcini and Cohen were jointly awarded the Nobel Prize for Medicine in 1986. Normally, one example of serendipity is enough to lead to a major scientific discovery. For the 1986 Nobel Prize, it required two.

A number of scientists knowledgeable in the field felt that Hamburger should have received a share of the 1986 Nobel Prize. It was in his laboratory that the discoveries were made; he was a joint author in many of the important papers; it was he who recruited both Cohen and Levi-Montalcini; it was he who provided funding and a suitable intellectual environment for them and it was he who did early groundbreaking work that led to the formulation of ideas about growth factors.

One final observation: Both Hamburger and Levi-Montalcini lived to celebrate their centenaries, the latter still alive and active at the age 102, with Dr Stanley Cohen a mere 89 at the time of this writing.

7 You Are What You Eat: How a Tooth Can Reveal Where You Came from and What You Ate

Who Are You/Where Did You Come from?

Mystery 1

A gruesome burial site was uncovered during road construction near Dorset, England in 2009. It contained the skeletons of 54 young men, with their heads separated from their bodies. There was no evidence of battle scars in other parts of the body, indicating that they had been executed. Radioactive carbon dating placed the deaths at about 1000 AD. No clothing or grave goods were found, so there was little information available as to who they were. Yet their identity is crucial in order to place this important burial in its correct historical context. Were they members of a local population or did they come from farther afield, and if so, from where?

Mystery 2

Stonehenge is one of the largest and most important Neolithic (new stone age) monuments in the world. Its circular arrangement of standing stones is sited in dramatic isolation on Salisbury Plain in Wiltshire, Southern England (Figure 7.1). It was constructed over a period of 1500 years, between BC 3100 and BC 1600. The effort and organisation required for its construction, involving shaping, transporting and setting up the large and heavy stones are a cause of wonderment. Arguments abound as to the purpose of the monument. Theories range from it being used as a temple for religious purposes, an observatory for astronomical observations, a calendar to help compute the seasons, a retreat for the sick to heal, a place to worship ancestors and a cemetery to bury the (important?) dead. The arrangement of the stones appears also to provide the site with special acoustic properties that can amplify sounds.

Although much is known about the physical aspects of Stonehenge, it is information derived from individuals found associated with it that gives it human interest and meaning. Great excitement always follows the discovery of a burial, especially if there are associated grave goods that may reveal features about the culture and status of the individual(s).

Nothing but the Tooth. DOI: http://dx.doi.org/10.1016/B978-0-12-397190-6.00007-9

Figure 7.1 The Neolithic site of Stonehenge in Wiltshire, England.

The richest burial to date was discovered in the village of Amesbury, about 5 km from Stonehenge. Like the Dorset burial site, the Amesbury burial was uncovered by accident when workers were laying the foundations for a new school. Dated to between BC 2400 and BC 2200, the grave contained the skeleton of a man about 40 years of age. Among the many valuable grave goods were 15 beautiful flint arrowheads (hence the nickname the 'Amesbury archer'), boar tusks, copper knives and gold hair-decorations (hence the alternative name 'King of Stonehenge'). Much information about the period has been derived from the origin of such grave goods, indicating the wide range of trade routes that existed at the time.

Despite all this treasure, perhaps the most important information that archaeologists wanted to know was where the archer came from. The simplest explanation would be that he was a chief who was born and lived locally. Alternatively, he may have come from Wales and been associated with moving the bluestones of Stonehenge that are known to have been quarried in the Preseli Hills, about 200 miles from Stonehenge. Could he have come from even farther afield? Each possibility would create a different personal history.

Another important burial near Stonehenge, on Boscombe Down, contained the skeleton of a 14–15-year-old boy, dated at about 800 years after the Amesbury archer. The grave contained a remarkable amber necklace with 90 beads (hence the nickname 'boy with the amber necklace'). However, no evidence was forthcoming as to where he came from.

Towards a Solution to Mysteries 1 and 2

Due to their durability, teeth and bones are usually the only remains preserved from the distant past. Rarely are soft tissues found, except for animals buried in frozen ground, such as mammoths. In the case of fish such as sharks and rays, which lack a bony skeleton (their bodies being supported by softer cartilage), only the teeth remain from extinct species, leaving much having to be surmised about their evolution by studying living species.

Examination of any skeleton can provide general information indicating the species, its age, sex and the likely nature of the diet (i.e. whether it was carnivorous, herbivorous or insectivorous; see Chapter 4). In human skulls, the presence of tooth decay would indicate a sugar-rich diet. The question arises as to whether these hard tissue remains can reveal more detailed information.

To solve mysteries 1 and 2, there needs to be a feature present in the skeleton that will indicate where it came from. Surprisingly, there is, and the secret lies in the water that an individual drinks during life, for the water carries a unique fingerprint that can pinpoint where the individual lived. To understand the science behind it, it is necessary to know a little about the chemistry of water and how elements in the water end up in the teeth and bones.

Oxygen in Water

Water is made up of a mixture of the elements hydrogen and oxygen, and its chemical formula is the well-known H_2O. This means each molecule of water has two atoms of hydrogen and one atom of oxygen. For our purposes, we need only consider the oxygen.

For any element, its atoms contain a number of subatomic particles in its central nucleus, including protons and neutrons. The number of proton particles is the same as the number of neutron particles. The two added together give the atomic number.

The main form of oxygen (O) has eight protons and eight neutrons, giving it an atomic number that can be written as ^{16}O (Figure 7.2). This form constitutes over 99.7% of the total oxygen. However, there also exists another stable form of oxygen that is present in minute amounts (less than 0.2%). This reacts in the same way, but because it contains two extra neutrons, it is therefore heavier and is known as ^{18}O (Figure 7.2). The different stable forms of an element are known as stable isotopes (as opposed to radioactive isotopes, which are unstable and break down; see page 110).

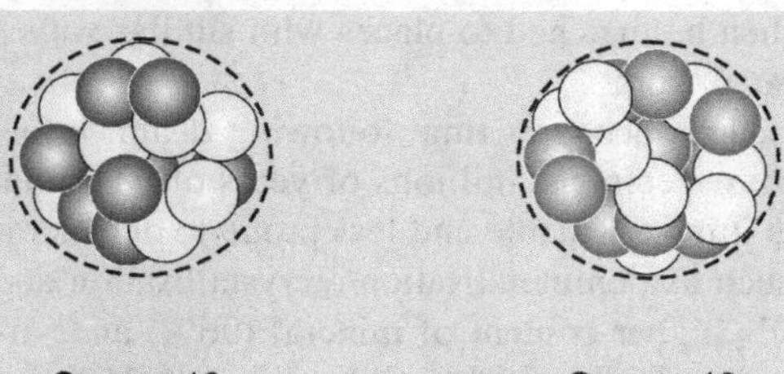

Figure 7.2 Diagram showing the nucleus of normal oxygen on the left (^{16}O) made up of eight protons (black) and eight neutrons (white). Its main stable isotope (^{18}O) on the right is made up of eight protons (black) and ten neutrons (white).

Drinking water is derived from water vapour evaporated off from the ocean, which falls from clouds as rain over the land. The heavier ^{18}O isotope will fall sooner and therefore be more common in the drinking water associated with a warm tropical climate nearer the ocean. The lighter, main form of oxygen (^{16}O) will be given off later at a further distance from the ocean and in colder climates towards the poles. The ratio of oxygen isotopes in the water varies according to latitude, water temperature and weather patterns. When someone drinks water, the oxygen isotopes in the water will be incorporated into their developing teeth and bones. Oxygen isotope levels for the majority of regions of the earth are known and are relatively stable with time (although corrective factors need to be applied to take account of climatic changes). By matching oxygen isotope levels in teeth and bones with the surrounding land, it is possible to determine where a person was born and lived. If there is a difference between oxygen isotope values of a specimen (especially when derived from enamel) and that of local specimens/ground where the skeleton was found, this implies migration from a different place of origin.

Structure and Composition of Teeth and Bones

Teeth and bones have two intermeshed components:

1. A hard *mineral* part containing the elements calcium, oxygen and carbon (together with trace amounts of strontium and lead).
2. A soft but strong organic *protein* framework collagen, present in bone and dentine, which contains the elements nitrogen and carbon derived from the food. Enamel differs from bone and dentine in that it contains virtually all mineral and no protein.

The different elements making up the teeth and bones are all derived from the food and fluid the individual consumes.

Distribution of Stable Oxygen Isotopes in Teeth and Bones

When an individual drinks water, the two forms of oxygen (^{16}O and ^{18}O) are incorporated into the structure of the mineral crystals (in both phosphate (PO_4) and carbonate (CO_3) components) during the formation of the teeth and bones. The oxygen can subsequently be released experimentally from the tooth using acids and then accurately measured in a piece of apparatus called a mass spectrometer. The ratios of the oxygen isotopes can then be matched to places with similar values to discover where the individual originated from.

Due to the stability of the mineralised tissues over time following death, information can be successfully obtained from specimens millions of years old. Enamel (see Figure 4.2) is particularly useful as it is very stable and less prone to degenerative changes after death (diagenesis), such as demineralisation/recrystallisation and loss of collagen. This is due to enamel's higher content of mineral (96%) and virtual lack of organic material (1% compared with 20% for dentine and 25% for bone). Bone, and to a lesser extent dentine, is more porous and more susceptible to change when buried in the ground following death due to its high content of collagen.

Solution to Mystery 1

With regard to the 54 skeletons uncovered in Dorset, at that time of burial England was being raided by the Vikings. In the case of the Dorset burial, analysis of the oxygen isotopes in the teeth showed conclusively that those young men were not locals but had originated from Scandinavia. The identity of the bodies as Viking warriors adds a whole new dimension to the context of the burial. This is one example where the generally successful Vikings came unstuck in their rape and pillage expeditions, and the whole group seems to have been captured and executed, presumably by the local Anglo-Saxons.

Solution to Mystery 2

When the enamel of the teeth of the 'Amesbury archer' ('King of Stonehenge') was analysed for oxygen isotopes, he was found not to be a local. In fact, his origin was much farther afield, from an Alpine region in central Europe that could be Switzerland, Germany or Austria. Although one mystery was solved, the information raised a host of new questions that may never be solved, such as why did he undertake such a long journey and why was he buried so close to such an important monument? When the new school near the site was opened in 2006, it was appropriately named the Amesbury Archer Primary School.

Oxygen isotope analysis from his teeth indicated that the boy with the amber necklace, like the archer, was not local but also originated from a considerable distance away. Unlike the Amesbury archer, who came from a cold, mountainous region, the boy came from a warmer, Mediterranean climate. These case histories indicate that, long into the distant past, Stonehenge already had an international clientele.

Having seen how the teeth and bones contain a unique fingerprint derived from drinking water that can tell where people (or animals) came from opens up the potential to solve many other important mysteries related to archaeology and evolution that cannot be answered by any other method.

Who Accompanied Christopher Columbus During His Second Voyage to America?

Isotope analysis of the teeth has the power to illuminate great historical events, one of which was the discovery (rediscovery?) of the New World of America by Christopher Columbus in 1492. This discovery also led to the establishment of the slave trade between America and Africa.

The skeletons of 20 individuals were recently unearthed at La Isabela, in the Dominican Republic, a place thought to be the first European town founded in the New World. Remarkably, there is strong evidence that these remains represent part of the original crew of Columbus's second visit to the New World (1493–1496). As historical records of the crew were incomplete, could isotope analysis of the teeth help indicate the birthplace of the crew? It is known that Columbus had a personal African slave on his voyages of discovery.

Stable isotope values for oxygen, together with those of carbon and strontium (see later in this chapter), from the enamel of the teeth provided a totally unexpected result. There was strong evidence that not one but three of the skeletons were of individuals born in Africa. This view was further supported by the findings that some of the teeth showed intentional dental modifications (whereby the teeth had their shape deliberately altered by filing; see Chapter 16) typical of those occurring in West Africa. These new findings indicate that more native Africans were involved in the first documented explorations of America than previously suspected.

What Can Teeth Tell Us about the Evolution of Elephants, Whales and Sea Cows?

As oxygen isotopes in teeth and bones provide information concerning water in the environment, can they add flesh to the bones by providing us with crucial evidence of an animal's environment that cannot be derived in any other way? For example, can it distinguish aquatic from land mammals? In the case of mammals such as whales, whose ancestors were land-dwelling creatures, can the teeth reveal when and who were the first ancestors that took to water?

In measuring stable oxygen isotope values in tooth enamel for different groups of animals, it has been determined that different values exist depending on whether the animal is aquatic (e.g. a whale), semi-aquatic (e.g. a hippopotamus) or terrestrial. Due to this important observation, data have been determined from the enamel in various fossils to determine whether they were aquatic, semi-aquatic or land based, with surprising results.

Elephants have previously always been considered land animals. Oxygen isotope values in the enamel of teeth were measured from two prehistoric elephants (Moeritherium and Barytherium) that lived over 37 million years ago. The results indicate that these ancestral elephants were not terrestrial but were semi-aquatic mammals, spending their days in water and feeding on freshwater plants.

Oxygen isotope values in enamel have also been investigated in 30–50-million-year-old fossils related to modern-day whales and sea cows (dugongs and manatees). The results indicated a different pattern in the evolution of the two groups. The early ancestors of whales were found to have inhabited freshwater environments, while their later ancestors moved rapidly into estuarine and marine environments. For the ancestors of sea cows, however, isotope values indicated an early transition to a marine environment without any evidence of an intermediate connection in freshwater habitats.

What Can Isotopes in Teeth Tell Us about the Life of Dinosaurs?

One group of animals that never fails to grip the imagination is the dinosaurs. These reptiles evolved 230 million years ago and were so successful that they were the dominant animals until about 65 million years ago, when the majority suddenly became extinct (except for modern birds now thought to be a form of dinosaur). The extinction is believed to have been related to a large asteroid slamming into the sea off Mexico's Yucatan Peninsula (Chicxulub crater) and also the volcanic eruptions associated with the Deccan Traps of central India.

Although much can be learnt from the skeletons of dinosaurs, some important questions concerning their biology and behaviour are still poorly understood. Three of these are:

1. Were dinosaurs cold or warm-blooded?
2. Did the large, herbivorous dinosaurs undergo an annual migration in search of enough food to live on (like living wildebeests today moving from the Masai Mara in Kenya to the Serengeti in Tanzania)?
3. How could two of the largest carnivorous dinosaurs, the spinosaurs and tyrannosaurs, coexist in the same territory (i.e. they were sympatric)?

Answers to these three problems have been sought by studying oxygen isotopes in the teeth.

When considering whether dinosaurs were warm-blooded (endothermic) like birds and mammals or cold-blooded (ectothermic) like modern-day reptiles such as crocodiles, morphological features such as the blood supply and structure of bone have been used to support the view that dinosaurs were warm-blooded. However, there is little in the way of any other hard evidence. Analysis of oxygen isotope values in the teeth has been employed to help answer this important question, as it is believed that these values will reflect environmental and possibly body temperature. Oxygen isotope ratios derived from mineral crystals in the teeth have been obtained for a wide range of cretaceous dinosaurs. These values have then been compared with those of crocodiles from a similar range of latitudes. The results clearly show consistent differences between the two groups. As crocodiles are cold-blooded, the results can only be interpreted as indicating that a wide range of dinosaurs were warm-blooded.

When considering the herds of massive, long-necked, herbivorous dinosaurs, the sauropods, it is very likely that they would have had to migrate to find enough food to satisfy their enormous requirements during the year. As an example of this group, the long-necked camarasaurids was chosen. This dinosaur lived about 150 million years ago, was about 18 m long and estimated to have weighed up to 30 T (although recent research suggests that dinosaurs may have been much lighter than was previously thought). To answer the question concerning migration, oxygen isotope values were measured from the enamel crystals in its teeth. If these dinosaurs did not migrate, then the oxygen isotope values would be the same throughout the full thickness of enamel and would reflect that of the surrounding environment. However, the isotope values showed differences with a regular periodicity across the enamel that could be matched with a region from highlands nearly 200 miles away, providing the first real evidence for migration of these large, herbivorous dinosaurs.

If large herbivores had to migrate considerable distances to obtain enough food during the year, it would imply that the carnivorous dinosaurs feeding on them would also have migrated. This question could be solved by similar studies using isotope analysis on teeth of such carnivorous dinosaurs.

In trying to understand how two of the largest carnivorous dinosaurs, the spinosaurs and tyrannosaurs, could have coexisted as top predators in the same territory, one clue is derived from the shape of their skulls and teeth. The skull and teeth of spinosaurs were crocodile-like, having an elongated snout and simple conical-shaped teeth, suggesting they primarily ate fish. The teeth of tyrannosaurs were massive and showed sharp, serrated edges adapted to meat-eating. Oxygen isotope studies carried out on the enamel of spinosaurs showed them to have values similar to those of coexisting, semi-aquatic crocodiles and turtles but significantly lower values than those associated with a variety of terrestrial-based dinosaurs. This evidence strongly suggests that spinosaurs were semi-aquatic. The semi-aquatic lifestyle correlates with their fish diet and to thermoregulation. Such a lifestyle meant that spinosaurs could have coexisted at the same time as the terrestrial-based tyrannosaurs, spending much of their time in the water eating fish and not competing directly with the land-based tyrannosaurs eating meat.

Where Did 'Otzi' Come from?

One of the major archaeological finds of recent years concerns the 'Iceman'. His well-preserved, mummified body was discovered in 1991 high up on an Alpine glacier in the Otztal Alps (hence his nickname 'Otzi') bordering Italy and Austria. It is estimated that he died about 5200 years ago, perhaps as the result of an arrow wound. His importance was due to the preservation in the cold temperatures of many artefacts such as clothes, tools and weapons. The physical presence of the Otztal Alps results in different isotopic oxygen profiles in water occurring north and south of it, the isotope values in the south being higher. Because Otzi was found on the border, there was a dispute between Austria and Italy as to which country the body belonged. With such a unique find, all the techniques of science were applied to discover as much as possible of the life and death of the 'Iceman'. Of prime interest was an answer to the questions of where he was born in relation to the place where died and was there any information as to his movements during life.

In addition to indicating the place of origin of an individual, looking carefully at the teeth and bones can also reveal important information as to the movements of an individual during life that is unobtainable by any other method. Although dependent on measuring oxygen isotope levels, this requires further understanding as to how teeth and bones form.

Bone is continually forming and being broken down throughout life (i.e. it undergoes remodelling). The more solid bone at the surface turns over slowly, and its stable isotope values reflect the origin of the individual in early life. The more delicate, spongy, central part of a bone turns over more rapidly, and its stable isotope values reflect the environment of more recent adult life.

Because different teeth develop at different times, enamel isotope values measured from different teeth indicate the environment at different periods during life. Thus, isotope values for a human first permanent molar, whose crown is formed by age 3, can be compared with values from a second permanent molar, whose crown is formed by age 7. Any difference will indicate a change in surroundings during childhood. Data from a third molar will give information on the environment in the early teen years. Taking very small samples across the full thickness of enamel from a single tooth that might take 3 years to mineralise can provide information about an individual's movement during those 3 years of childhood. Dentine, being slowly deposited throughout life, can provide information more related to an individual's environment after the age of about 20 years.

Analysis of oxygen isotopes (as well as those of strontium; see later in this chapter) from both enamel and bone samples solved the question of whether the Iceman was born in Austria or Italy. He had in fact lived on the Italian south side and spent his early childhood in a valley region within a range of about 60 km from the site where the body was found. In addition, different oxygen isotope values later in life from bone indicated migration to higher ground.

What Did You Eat?

The availability and the type of food have driven evolution. Animals that have specialised in eating one type of food may be very successful while that food is plentiful, but if it becomes scarce as a result of, say, environmental changes, they may not be able to adapt successfully to another type of food and may become extinct. During the evolution of any species, there are numerous fossils in the family tree that may have fed on a diet different to that of related forms today. The question is: Is there anything in the teeth and bones of a fossil that may indicate the nature of the diet? Consider the following two examples.

Mystery 3

During man's evolution, many early bipedal (walking upright on two legs) ancestors evolved and separated off from the line that led to our closest living relatives, the great apes (see Chapter 13). Is there any information about diet that can be discovered by studying the skeletal remains of our ancestors that might help explain why some survived and others became extinct, and how their diet differed from that of modern great apes? The largest of all the robust australopithecines, namely *Paranthropus boisei* (also known as *Zinjanthropus boisei*; see Chapter 13) was nicknamed 'nutcracker man' because of his huge grinding molars (Figure 7.3). It has been assumed that the large teeth were used to deal with hard food items such as nuts, fruit and seeds. Is this true?

Mystery 4

One of the key events in establishing the early civilisations in America was the development of agriculture, particularly the cultivation of maize. Is there anything

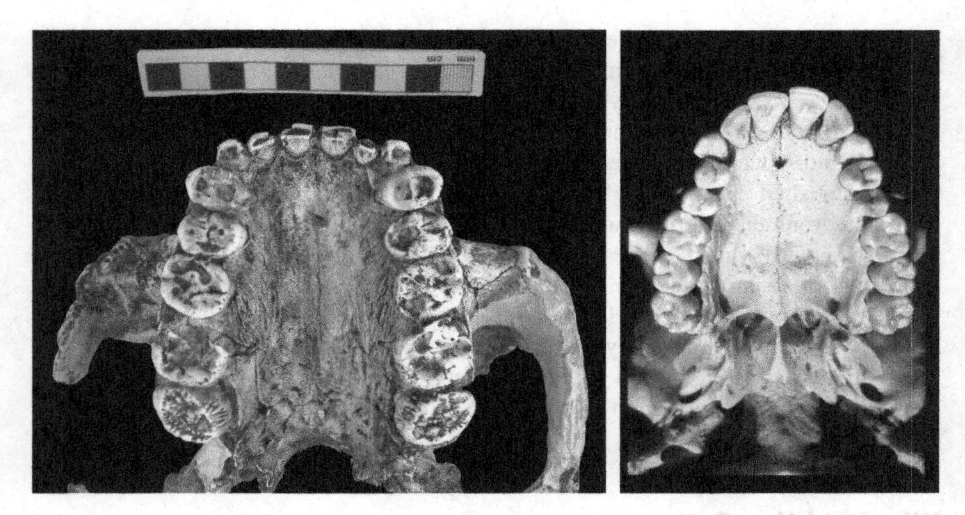

Figure 7.3 Upper jaw of *Paranthropus boisei* (left) with its very large molar teeth compared with those of modern humans (right).
Source: Courtesy of Professor C. Dean. Photographed by M. Farrell.

in skeletal remains found during this period that can specifically detect maize in the diet?

Towards a Solution to Mysteries 3 and 4

The solution to identifying the diet of an individual again lies in studying isotopes present in the teeth and bones. The methods are basically similar to those discussed previously using oxygen isotopes. However, in this instance the isotope involved is carbon (and also nitrogen; see later in this chapter), and it is first necessary to understand how carbon gets into the food chain and then into the teeth.

Carbon in the Food Chain

Carbon (C) exists in the atmosphere as carbon dioxide, each molecule consisting of one carbon atom linked to two oxygen atoms (CO_2). Carbon dioxide and water are taken up by plants and algae and during photosynthesis are converted into sugar for fuel. The green pigment, chlorophyll, is very important in this chemical reaction as it allows plants to absorb energy from sunlight. Plants (and algae) lie at the base of the food chain, and all life depends on them.

Stable Carbon Isotopes

The element carbon (C) has six protons and six neutrons, giving it an atomic number of 12, which is written as ^{12}C (Figure 7.4). It comprises 99% of carbon.

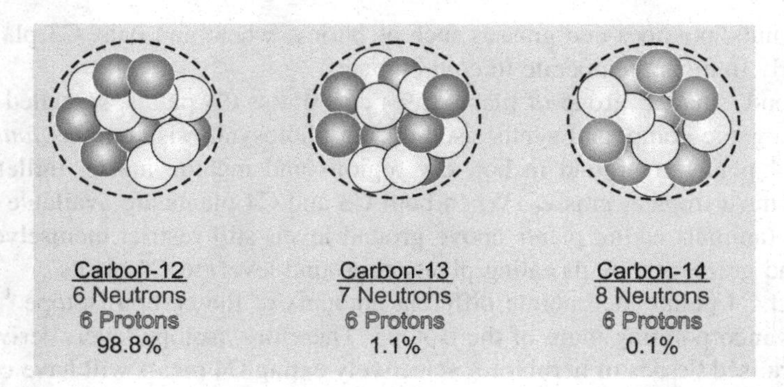

Figure 7.4 Diagram showing the nucleus of normal carbon on the left (^{12}C) made up of six protons (black) and six neutrons (white). In the middle is its main stable isotope (^{13}C) made up of six protons (black) and seven neutrons (white). On the right is the unstable radioactive form (^{14}C) with six protons (black) and eight neutrons (white).

However, very small amounts of carbon (approximately 1%) exist in a different, but stable, form and react in an identical manner, the only difference being that it has an extra neutron in the nucleus and is therefore slightly heavier. This stable isotope is written as ^{13}C (Figure 7.4), and it is possible to measure the amounts of carbon isotopes incorporated into plants during photosynthesis.

During photosynthesis, the lighter, main form of carbon, ^{12}C, is enriched. When the plant is eaten by an animal or human, the digestive process causes a further change, resulting in an increase in the heavier isotope, which is incorporated into the teeth and bones of the animal. However, the ratios of the two stable isotopes of carbon will differ according to the nature of the diet. Research using live animals on controlled diets has shown that there is a characteristic carbon isotope value for animals eating a plant diet (herbivores), eating meat (carnivores) and eating both types of food (omnivores).

There is even more information available about diet as carbon isotope values for marine plants differ from those of terrestrial plants. Marine herbivores, such as the manatee, will have isotope levels that distinguish them from land herbivores. Carnivores feeding on the meat of marine herbivores will also have isotope levels distinguishing them from carnivores feeding on land herbivores.

C3 and C4 Plants

The analysis of carbon isotopes from animal skeletons can inform us not only that an animal is a herbivore or carnivore and the percentage of the diet that is plant-related, but also it can provide information on the type of plants eaten.

In the manner of their photosynthesis, plants can be divided into two main groups. The majority of plants (95%) are known as C3 plants, as the first organic compound synthesised from carbon dioxide in the air during photosynthesis contains *three* carbon atoms. These C3 plants comprise trees, shrubs, flowering plants, fruit and nuts, potatoes and grasses such as barley, wheat and oats. C3 plants exist particularly in moist, temperate to cold climates.

A second, smaller, group of plants (5%) constitutes C4 plants, so-called because the first organic compound synthesised during photosynthesis contains *four* carbon atoms. C4 plants are found in hot, dry regions and include maize, millets, sugar cane and most tropical grasses. When both C3 and C4 plants are available as food, browsers (animals eating plants above ground level) still restrict themselves to C3 plants, and grazers (animals eating plants at ground level) to C4 plants.

C3 and C4 plants incorporate different amounts of the carbon isotope ^{13}C, with C4 plants incorporating more of the isotope. Therefore, isotope levels derived from the mineralised tissues of herbivores selectively eating C4 plants will have comparatively more of the isotope than herbivores selectively eating C3 plants. Intermediate levels would indicate a herbivore with a diet incorporating both types of plants. In addition, carnivores eating the meat of land herbivores will have yet another distinctive isotope value. As well as indicating the type of plants eaten by herbivores, the identification of C4 plants in a diet would show the existence of a dry, arid climate.

Distribution of Carbon in Teeth and Bones

Carbon exists in teeth and bones in two situations. In the mineral crystals, it is part of calcium carbonate ($CaCO_3$), while in bone and dentine, it also forms part of the organic collagen content. Enamel is particularly useful as a source of carbon isotopes as enamel is virtually all mineral and undergoes the least change of all the skeletal tissues following death.

Solution to Mystery 3

The application of carbon isotope ratios has helped to provide answers to some of the questions raised in mystery 3. Our closest living relatives the chimpanzees and gorillas, eat an almost pure C3 diet of fruit and leaves. One of our early ancestors in Africa, *Ardipithecus ramadus*, lived about 4.4 million years ago and, although having a small brain, walked upright. It retained a prehensile (grasping) big toe, suggesting it climbed trees and lived in a wooded environment. Carbon isotope levels from its enamel indicate that, although it existed mainly on C3 plants, its diet was more varied than that of great apes, consuming between 10% and 25% of C4 plants (e.g. grasses). This contrasts with that of the later, more advanced group of bipedal hominins, the australopithecines, where carbon isotope values indicate that the diet was made up of more than 30% C4 plants (with variation ranging up to 80%).

The largest of the robust australopithecines, *Paranthropus boisei*, has now been found not to have used its huge grinding molars (Figure 7.3) to deal with hard C3 food items such as nuts, fruit and seeds, as was originally thought. Using carbon isotope analysis from the enamel of more than 20 individual *Paranthropus boisei* teeth 1.9–1.4 million years of age, it has been discovered that, quite unexpectedly, the diet was totally dominated by C4 material. Its large teeth and powerful masticatory system were therefore used to break down tough but not very energy-rich food, such as grasses or sedges.

The 'bushiness' of the human family evolutionary tree is evident by the considerable number of early ancestral forms found, including a number of different species of australopithecines. The remains of a new species named *Australopithecus sediba*, including two very well preserved individuals, were recently found in 2008 in Malapa, South Africa, and dated to around 2 million years ago. They represented a male child in his early teens and a female thought to be roughly 30 years old. Because of the presence of both primitive and more advanced features, researchers have suggested that *Australopithecus sediba* may be transitional between *Australopithecus africanus* (e. g. the Taung child; see Chapter 13) and either *Homo habilis* or the later *Homo erectus*. Studies using carbon isotope analysis of the enamel showed the former's diet differed from that of *Ardipithecus ramadus* and *Paranthropus boisei*, being almost exclusively C3, despite the availability of C4 food. This was confirmed by studying the unusual presence and retention of dental calculus (tartar) around the teeth. From this material, researchers were able to isolate and identity phytoliths (plant-produced silica bodies), which have characteristic shapes that determine the specific plant. The phytoliths present again demonstrated that the diet was exclusively C3 plant material, such as fruit and leaves, and included wood and bark. Their diet was therefore more chimp-like.

Carbon isotope studies have therefore shown that man's potential early ancestors appear to have had a variety of diets, and those on the pathway to our own species would likely demonstrate a more varied diet and exploit a wider range of woodland resources than some of the australopithecines and modern-day chimpanzees and gorillas.

Solution to Mystery 4

Maize (corn) was domesticated about 10,000 years ago in Mexico. It appears to have been selected, cultivated and enormously improved nutritionally from the wild grass, teosinte. The change from small populations of mobile, hunter-gatherers to a sedentary, agricultural society was a major event in the history of the American continent. Once agriculture was firmly established, it produced an increasing supply of food that allowed for the development of greater populations and eventually civilisations, such as the Maya in Mexico. As maize is a C4 plant, its establishment and spread throughout America has been traced by studying stable carbon isotopes in the teeth and bones found at burial sites. This has shown that maize production spread from Mexico over the whole of the American continent, both North and South. When the Spanish conquistadors arrived in the New World at the beginning of the sixteenth century, they transported maize back with them to Europe. Also, knowing the status of an individual skeleton from the presence and nature of burial goods, it has been possible to show that there were differences in the consumption of maize within the social hierarchy.

Analysis of carbon isotopes in teeth have helped to answer many questions related to archaeology and evolution that cannot be answered by any other method, including the following.

What Can Carbon Isotope Analysis Tell Us about the Evolution of the Large African Herbivores?

Methods of distinguishing between C3 and C4 plants using carbon isotope data have enabled scientists to probe evolutionary aspects associated with Africa's great herbivorous mammals. C4 grasses evolved in Africa between 15 and 10 million years ago. Before this, herbivores all browsed on the available C3 trees and shrubs. By analysing the enamel in the teeth of different herbivore species, it has been possible to determine when species changed from a C3 browser diet to a C4 grazer diet. For example, the ancestors of zebras were among the first to incorporate the new C4 grasses into their diets between 9.9 and 7.4 million years ago, whereas the ancestors of warthogs did not incorporate C4 grasses until much later (6.5–4.2 million years ago).

Were Dinosaurs Warm-Blooded?

In support of the view that many dinosaurs were warm-blooded, it was stated on page 99 that their oxygen isotope levels were different from those of cold-blooded dinosaurs. Is there any more supportive isotope evidence about warm bloodedness in dinosaurs?

Although carbon and oxygen isotope analyses are often undertaken separately, a new methodology combining both these isotopes has been recently introduced. This is based on the hypothesis that, in the carbonate (CO_3) component of the mineral crystals of teeth, the trace amounts of the carbon isotope ^{13}C will combine directly with the trace amounts of the oxygen isotope ^{18}O, and that the warmer the temperature of the animal as teeth form throughout life, the less frequently will the two isotopes combine. Important evidence to support this interpretation was provided when it was shown that values for the combination of the carbon and oxygen isotopes in enamel were consistently higher in living, cold-blooded reptiles such as the crocodile, compared with warm-blooded mammals such as the elephant and rhinoceros. As enamel is so stable over time, the same methodology has been applied to enamel in the teeth of the huge, long-necked, sauropod dinosaurs living well over 100 million years ago, to determine whether these extinct reptiles were cold-blooded or warm-blooded. The results indicate that body temperatures in these dinosaurs were between 36°C and 38°C, which is similar to those of most modern mammals.

Did the Diet of Neanderthals Differ from that of Early Humans Living at the Same Time?

Fifty thousand years ago, two human species coexisted, Neanderthal man and true *Homo sapiens*. Why did we survive and the Neanderthals did not? The same food in the form of fruit, nuts and meat presumably was available to both, but did both make full use of it? Was one a better hunter than the other and more readily able to obtain meat, even if times were hard? Is there any information derived from skeletons that can help answer these questions?

A comparative study measuring carbon isotope levels in the teeth of Neanderthals and very early specimens of our own species has provided some information towards answering this question. However, one other important source of food not readily identified by carbon isotopes is fish and other types of marine food. This type of diet is more readily identified using a different stable isotope, namely the element nitrogen.

Stable Nitrogen Isotopes

The main principles laid out for stable carbon and oxygen isotopes also apply to the element nitrogen (N). Its most abundant form is written as ^{14}N and represents about 99.6% of nitrogen. The nucleus has seven protons and seven neutrons. Nitrogen has an additional stable isotope known as ^{15}N that has an extra neutron and is present in trace amounts (less than 0.4%).

Nitrogen is absorbed by land plants from the soil. When these plants are eaten by land herbivores, the nitrogen is incorporated into the collagen protein making up a considerable component of bone and dentine (and other soft tissues). Like carbon, the $^{15}N/^{14}N$ ratio increases as it passes up the food chain (fractionation or trophic enrichment). Nitrogen's trophic enrichment is even greater than carbon's. Detailed research has shown that isotopic analysis of stable nitrogen isotopes in the collagen of mineralised tissues, as already seen for carbon, gives a characteristic value in animals eating a chiefly plant diet (herbivores). Carnivores that eat land herbivores will have an even higher ratio of nitrogen isotopes, while omnivores with a more varied diet would have an intermediate value between that of carnivores and herbivores. However, an important feature of nitrogen isotopes is that they allow for a good distinction between terrestrial and marine environments, as marine animals have a higher concentration of ^{15}N.

Nitrogen isotope studies are therefore particularly informative where the diet includes fish, as animals eating fish or marine mammals will have even higher values than those preying on terrestrial sources. Individuals consuming a mixture of terrestrial and marine foods will have intermediate nitrogen isotope values.

Stable isotope values for nitrogen as well as for carbon have been determined from collagen in the bones of Neanderthals that lived between 120,000 and 27,000 years ago. These values have then been compared with those obtained for the bones of early modern humans living between 40,000 and 27,000 years ago. The results indicate that Neanderthals obtained all, or most of, their protein by consuming the flesh of herbivores. On the other hand, early modern humans showed more variety in their diet, also eating fish and other marine species. This adaptability to utilise other food sources could have been important in the struggle for survival between the groups.

Strontium and Radioactive Carbon

Two other elements occurring in the teeth and bones have also been employed to provide important information in the field of archaeology and evolution. These are strontium and radioactive carbon and both will be briefly considered.

Strontium

The element strontium (Sr) has two isotopes that, like those of oxygen, can help tell where an individual was born and lived. These isotopes are written as ^{86}Sr (lacking one neutron) that accounts for about 10% of the element, and ^{87}Sr (lacking two neutrons) that accounts for 7%. ^{87}Sr does not occur naturally but is formed by radioactive decay of another element, rubidium. It therefore accumulates slowly through geological time. In rocks that are old (more than100 million years), the $^{87}Sr/^{86}Sr$ isotope ratio is high. In younger rocks (less than100 million years), the $^{87}Sr/^{86}Sr$ ratio is low. The ratio of these two isotopes, $^{87}Sr/^{86}Sr$, provides a unique 'fingerprint' to the geology of an area.

Like carbon and nitrogen, strontium isotopes make their way from the soil into plants and thence up the food chain, reflecting the local geological environment. As it is similar to calcium, strontium readily substitutes for calcium in the inorganic crystals of mineralised tissues and can attain relatively high concentrations (100 parts per million). The most important use of strontium isotope levels from teeth and bones relates to the fact that they reflect the geology of the region where the skeleton came from. Global maps exist for the ratio of strontium isotopes occurring throughout the world. By matching strontium isotope values found in the mineralised tissue of an individual with the surrounding land, it is possible to deduce the place of origin of the individual.

Mystery 5

The Inca rulers in Peru were known for carrying out human sacrifice, including children, to appease their gods. A well-furnished, recently discovered grave containing the remains of seven children aged between 3 and 12 years is believed to represent the remains of sacrificial victims. It has been surmised that children in such a situation were gathered together from different regions of the Inca Empire. Can examination of the teeth provide any evidence to support this theory?

Solution to Mystery 5

Strontium isotope analyses of enamel from the teeth of the children were measured to see whether they had been obtained from different parts of the kingdom. The results supported the hypothesis in that, while five of the bodies represented children born and raised in the local region, two of the bodies were those of non-locals.

Like the Incas, the Mayan civilisation in Mexico also carried out human sacrifice. During excavations in one of the largest pyramids (the Moon Pyramid) in Teotihuacan, one of the great Mayan cities during the first millennium, a number of apparent sacrificial victims were uncovered, dating to a period about 150–350 AD. The results from strontium isotope analysis on the teeth of these victims indicated that they were not all local but had been gathered from different parts of the surrounding countryside.

Mystery 6

Slave burial grounds have been unearthed in Campeche, Mexico (dating from between the mid-sixteenth and late-seventeenth centuries), and in Rio de Janeiro in Brazil (dating from between 1760 and 1830). Due to the limited documentation of the early slave trade, could any information be found as to where the slaves originated? Had they been captured in Africa and transported to Mexico and Brazil or had they been born in these two countries?

Solution to Mystery 6

Enamel from the permanent teeth of 10 of the specimens at the Mexican site was analysed for strontium isotopes, and results showed that while six had been born locally, four of the subjects had been born in Africa and transported to Mexico. In the case of the specimens from Brazil, however, strontium isotope analysis has shown they were all born in Africa. Furthermore, instead of being limited to the Western coastal region of Africa as once thought, they reflected a much wider origin involving the East coast and the Central regions. In support of their African origin, a number of the teeth showed intentional tooth modification characteristic of that region (see Chapter 16).

Evidence for Migration and Social Customs

Two further examples are given to show how the use of strontium isotope analysis on teeth has provided information on the movements of individuals and on social customs.

A third important burial site near Stonehenge was found at Boscombe Down. This burial was unearthed in 2003 when workmen were digging a trench for a new water pipe. Although most Neolithic burials contain a single individual, this site was unusual as it contained the remains of seven individuals: three male adults, one male teenager and three children. There were also a number of grave goods, including pottery, a boar's tooth and five flint arrowheads, giving the group the nickname the 'Boscombe Bowmen'. To obtain more information about the origin and movements of the adults, combined strontium and oxygen isotope analyses were undertaken on the enamel of premolar teeth and third permanent molars. The premolars would provide information as to their whereabouts between the ages of 3 and 6 years and the third molars their whereabouts between 9 and 13 years. The results provided three findings of interest. First, there was a difference between the readings taken from the two teeth, indicating that they had lived in two different areas during the two time periods. Second, the region which best matched the isotope values was the area in Wales from which the bluestones of Stonehenge were quarried. Third, the values also differed from the site of their burial at Stonehenge, indicating that they moved among at least three different areas during their lives. Isotope levels in two of the children buried in the same grave indicated that they did not share the same homeland as the adults.

As strontium isotope analysis can provide information about where someone is born, it can also throw light on cultural behaviour in the past. A group of skeletons 4600 years old was found recently near Eulau, Germany, and has provided an unusual glimpse into possible marriage customs in the region at the time. They consisted of men, women and children who suffered a violent death simultaneously and probably represented related family members. Analyses of strontium isotope ratios from their molar teeth indicated that the men and the children were locals who lived and died in the Eulau region. The women, however, were brought up at a different location from where they raised their children, the closest location with similar isotope values being 60 km away. These results can be interpreted to indicate that all the women came as brides from a different community (exogamy) and went to live with the families of their husbands (patrilocality) in this early Neolithic community.

Radioactive Carbon

In addition to its stable isotopes ^{12}C and ^{13}C, carbon is present in a third form, namely the radioactive isotope ^{14}C. This isotope contains six protons and eight neutrons (Figure 7.4) and it is unstable, eventually decaying into nitrogen 14 (^{14}N). It comprises an infinitesimal amount of carbon (one part in a trillion!) but this is measurable using very sensitive apparatus. As its decay rate in organic material is known, it is used in radiocarbon dating, whereby the age of any archaeological material containing carbon can be accurately assessed up to about 60,000 years ago.

The main source of ^{14}C is cosmic ray action on nitrogen in our atmosphere. Additional contributions also appear as a byproduct from the explosion of atomic bombs. Although only two bombs were dropped on Japan at the end of the Second World War in 1945, many bombs were exploded during the testing of nuclear weapons between 1955 and 1963, and before a nuclear test ban treaty was signed. Radioactive carbon, like normal stable carbon isotopes, is taken up by plants in the form of carbon dioxide and, through this pathway, is incorporated into developing enamel when people (and animals) eat the plants. The level of ^{14}C present in enamel reflects the level in the atmosphere at that time. By measuring the ^{14}C levels in enamel and comparing them with the known records of ^{14}C in the atmosphere, it is possible to age a person born after 1953 to within about 1.6 years, just from an isolated tooth. This method is more accurate than other methods for aging adults. An absence of any significant radioactive ^{14}C isotope would indicate a person born before 1945. The technique has been used to help in the identification of some victims of the tsunami in Southeast Asia in 2004.

Summary

From this chapter's examples, it can be seen that detailed chemical analysis of mineralised tissues, especially teeth, has revolutionised our knowledge of the past.

It can tell us where an individual or animal came from, whether it migrated to another area later in life and what its diet was. It can even inform us as to whether an animal is cold-blooded or warm-blooded and has provided the correct historical context of many important burials. Now, whenever important skeletal material is discovered, the first thought of the researcher is to wonder what story the teeth will tell. It supports the contention that ‘you are what you eat’.

8 Teeth of Rock-Climbing Gobies: The Most Remarkable Dentitions in the Animal Kingdom?

A strong candidate for the most remarkable dentition in the whole of the animal kingdom is a group of fish, the rock-climbing gobies. Typified by the monk goby (subfamily *Sicydiinae*, including *Sicyopterus japonicus*), these are small undistinguished-looking fish reaching a length of about 10 cm (Figure 8.1).

Rock-climbing gobies spawn in the fresh water of fast-flowing rivers, and their newly hatched larvae drift down to the sea. Here they pass through a larval stage lacking teeth and feed on microscopic plants and animals in the plankton. The post-larvae return from the sea some months later to start their journey upstream in freshwater rivers.

Apart from the physiological challenge of accommodating to a freshwater environment following a marine one, the fish now have to adapt to two other major changes. The first is that the fast-flowing rivers they inhabit in Far East countries like Japan, Indonesia and Hawaii often contain waterfalls that they need to climb. The second is that their diet changes and they feed off algae that grow on stones on the river bed.

To move upstream, by the time they arrive at the mouths of estuaries the rock-climbing gobies have developed a powerful sucker on their under surface (by fusion of their two pelvic fins) in order to counteract the tumbling flow of water in the opposite direction.

The adaptation to feeding on algae-colonising stones on the river bed requires an amazing metamorphosis of the head region as the gobies start to make their way upstream. Firstly, from its position at the very front of the head, the mouth is relocated to a situation where it lies underneath the head, like a shark or ray, so that when the goby attaches to a rock with its ventral sucker, its mouth is favourably positioned to be in direct contact with the algae-covered rock on which it feeds. Secondly, the upper lip enlarges to form another sucker at the front of the head. Thirdly, teeth appear inside the mouth to enable the goby to scrape the algae off the rocks. These dramatic changes in the morphology of rock-climbing gobies take place very rapidly, probably in terms of just a few days.

When it encounters a waterfall (Figure 8.2), the rock-climbing monk goby scales it by alternately attaching to the rock via its mouth sucker and chest sucker, aided by muscular contractions of its body (Figure 8.3). This allows the fish to gain access to the water immediately behind the top edge of the waterfall, a habitat denied to other fish.

Nothing but the Tooth. DOI: http://dx.doi.org/10.1016/B978-0-12-397190-6.00008-0

Figure 8.1 Picture of a rock-climbing monk goby (*Sicyopterus japonicus*). *Source*: Courtesy of Dr N. Sahara.

Figure 8.2 A waterfall climbed by rock-climbing gobies. *Source*: Courtesy of Mr N. Fukuchi.

In considering the teeth in more detail, rock-climbing gobies are unusual in that there are major differences between the teeth of the upper and lower jaws. While there are only a few, spaced, conical teeth in the lower jaw, those in the upper jaw form a single, continuous row of up to 60 or more tiny teeth on each side (Figure 8.4). Unlike the simple, conical shape of the lower teeth, the upper teeth when they first erupt have a more complex shape, being surmounted by three small cusps (Figure 8.5). The teeth are slightly hinged at their attachment to the bone of

Figure 8.3 High-power view of rock-climbing gobies scaling the waterfall shown in Figure 8.2.
Source: Courtesy of Mr N. Fukuchi.

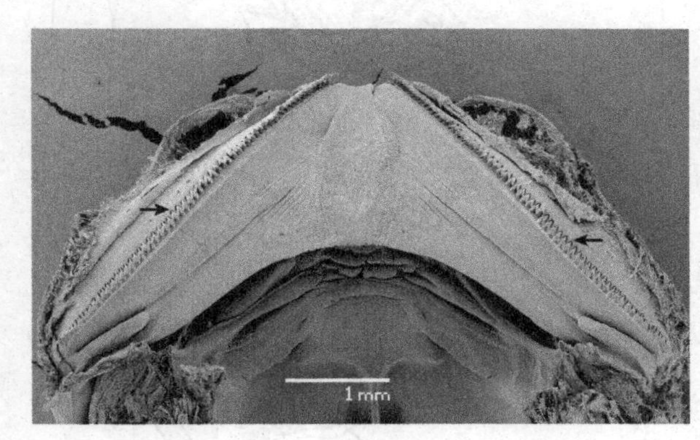

Figure 8.4 Illustration showing a row comprising 50–60 erupted teeth on each side of the upper jaw of a rock-climbing monk goby (arrows).
Source: From K. Moriyama et al., 2009. Courtesy of the editors of the *Journal of Oral Biosciences*.

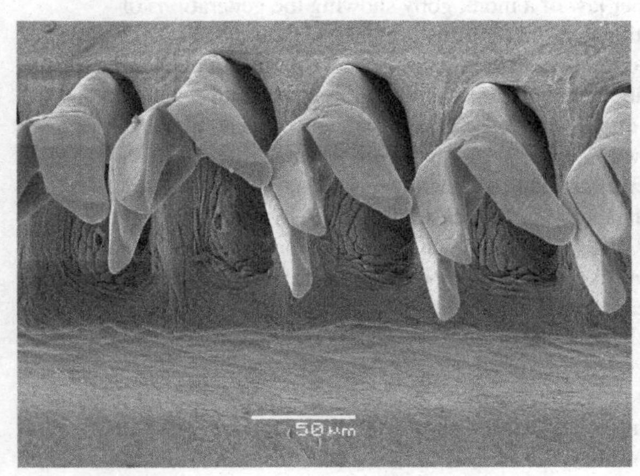

Figure 8.5 Upper teeth of a rock-climbing monk goby showing the presence of three small cusps when newly erupted.
Source: From K. Moriyama et al., 2009. Courtesy of the editors of the *Journal of Oral Biosciences*.

the jaw, giving them a small degree of flexibility, which is useful in helping the teeth as scrapers.

It is when studying the teeth in the upper jaw that the unique specialisation of rock-climbing gobies becomes apparent. Tooth replacement is considered in more detail in Chapter 9, where it will be seen that such replacement occurs throughout life in non-mammalian vertebrates (fish, amphibians and reptiles). Regardless of age, these animals will always show the presence of one or two sets of replacing teeth. In the exceptional case of the sharks and rays, up to about 10 replacing sets are evident in the jaw (Figure 9.1). This specialisation reflects the fact that teeth in the shark are being shed and replaced rapidly every few weeks.

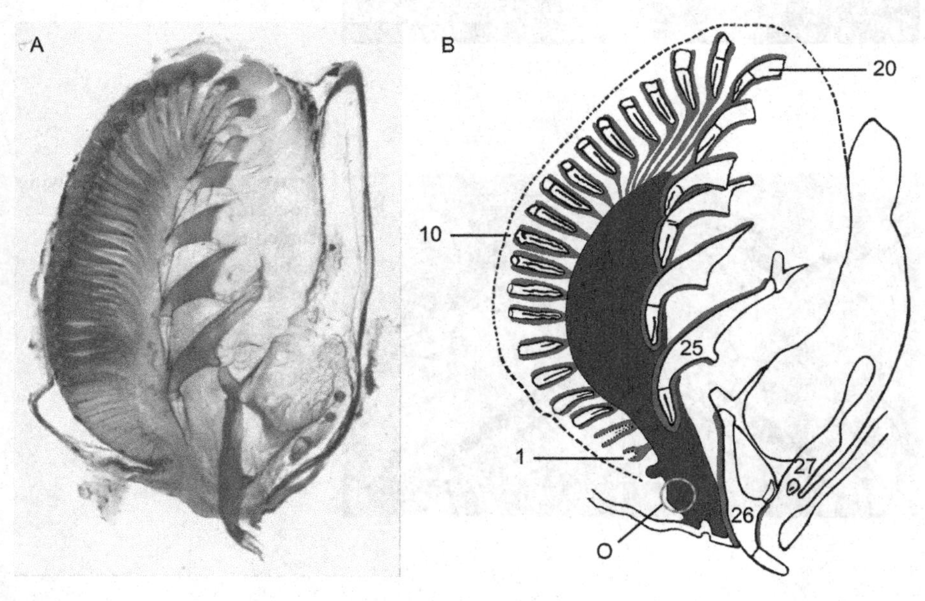

Figure 8.6 (A) Section of the upper jaw of a monk goby showing the generations of replacing teeth beneath each functional tooth. (B) Line drawing of Figure 8.6A. In this 3-cm specimen of a monk goby, a total of 26 generations of teeth are visible from the youngest labelled number 1 to the oldest (labelled number 26) which has erupted into the mouth and becomes attached to the jaw. The open circle represents the site at which all new teeth form from the shaded zone. As each new tooth starts to grow, it moves upwards. The 10th (10) and 20th (20) generation of teeth are labelled, while the 25th generation tooth (25) is almost complete and lies immediately beneath the functional tooth (26). As soon as the functional tooth is lost, it will be replaced by tooth number 25 beneath it. All the other teeth will move up one position and a new tooth will be budded off beneath tooth 1, which will then become tooth number 2. The remnants of a previously erupted tooth labelled number 27 can be seen enclosed by the lining of the mouth. An older, 10-cm specimen can have 45 generations of teeth.
Source: (A) from K. Moriyama et al., 2009. Courtesy of the editors of the *Journal of Oral Biosciences*. (B) from K. Moriyama et al., 2010. Courtesy of the editors of *Cell Tissue Research*.

Instead of possessing the relatively large number of 10 sets of replacing teeth, imagine an animal with 20 sets of replacing teeth present at any one time. In a 3-cm-long, rock-climbing goby, up to 25 sets of replacing teeth may be present. In a larger, 10 cm specimen, up to 45 sets of replacing teeth can be present beneath each of the 60 erupted teeth in the upper jaw, giving a total tooth count of about 2700 in its small head!

Each replacing set of teeth is arranged in the form of a compressed circle, with each individual replacing tooth being at a progressively younger stage of development than the one above it. This is shown in Figure 8.6A and B, where a rock-climbing monk goby is illustrated with 26 sets of replacing teeth. The youngest tooth is at the base of the first arm of the 'n', and the oldest and sole functioning tooth lies at the base of the second arm of the 'n'. Whenever this functional tooth is shed, all the replacing sets move up one position along the compressed circle and a new tooth is budded off in the deepest position. The effect resembles an escalator. From Figure 8.6, it can be seen that a base to the crown of the tooth develops comparatively late (from about tooth generation 18–26). This base will quickly fuse to the jaw only as the tooth becomes functional.

Another unique feature of rock-climbing gobies is that, whenever a tooth is about to be replaced, instead of being completely shed into the surrounding water, most of it is resorbed beneath the surface of the mouth (Figures 8.6 and 8.7). Presumably this allows the fish to reutilise important nutrients, such as calcium, in the teeth.

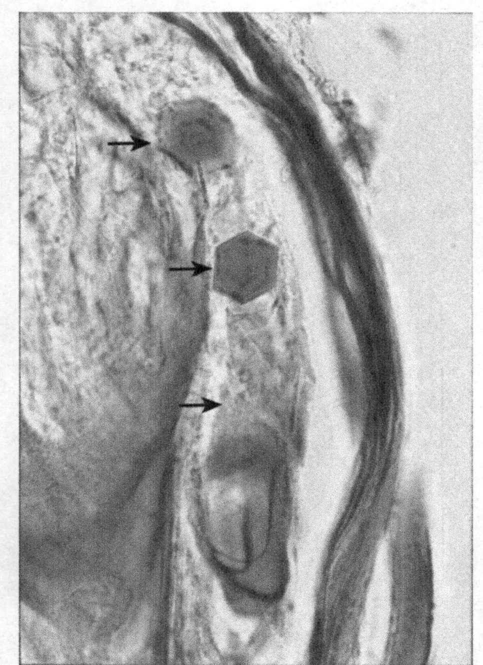

Figure 8.7 High-power view of inset in the region near the base of the functional tooth labelled number 27 in Figure 8.6B. It shows evidence of fragments of older functional teeth (arrows) that have been engulfed and are being digested, perhaps with the aim of reutilising their important constituents such as calcium.
Source: Courtesy of Dr N. Sahara.

The uniquely large number of replacing sets of teeth in the rock-climbing monk goby must indicate that tooth replacement is extremely rapid, almost certainly the fastest in the animal kingdom. It is estimated that each tooth is replaced after only 9 days! The reason for this remarkably fast rate must be because the effort in scraping algae from the surface of stones results in rapid wear and flattening of the cusp tips. Worn teeth must be less efficient at scraping the algae off stones compared with the newly erupted three-cusped form and need to be replaced.

So, if ever you meet a rock-climbing goby, pay it the respect it deserves: in its small, 1 cm-long head, it will have more teeth than the biggest great white shark.

9 Why Can't I Have Lots of Sets of Teeth Like a Shark?

When a child's milk tooth drops out, it is put under the pillow overnight for parents to substitute money on behalf of the tooth fairy. It must be quite expensive for the poor tooth fairy, but she is lucky the child is not a shark, as that would cost her a whole lot more, because sharks have many more teeth than humans, which are shed and replaced continuously throughout their lives. Why can't we have lots of sets of teeth, like a shark? Although this might seem a wonderful idea to people who have lost teeth to decay or gum disease, it really isn't.

The reason why the teeth of fish, amphibians and reptiles are continuously replaced is that these animals grow throughout life, and the size of their teeth has to keep pace with their growing jaws. Small teeth may be fine for a small shark but would be of little use when it has grown into a large one. The main differences between the teeth of these types of animals and those of humans have been mentioned in Chapter 4 (page 55). Humans have teeth with different and more complex shapes, including broad (molar) teeth that occlude and grind up the food to gain quicker access to its energy. This requires careful positioning of the teeth and hence roots are fixed to the jaw by a very slightly mobile fibrous joint. Grinding also calls for a wider range of jaw movements carried out by a more complex muscle system, compared with the simpler opening and closing movements found in the non-mammalian vertebrates. Because of these features, mammals have evolved a different solution to the problem of increasing size compared with sharks.

Tooth Replacement in Fish, Amphibians and Reptiles

The teeth of sharks and other non-mammalian vertebrates (fish, amphibians and reptiles) have a simple shape and do not have roots. The teeth in opposing jaws do not meet, so that rarely can the food be cut up or crushed into smaller pieces. The function of the teeth is merely to help grip the food and stop it escaping before swallowing it whole.

To maintain the right size of tooth for the enlarging jaws, tooth replacement is continuous throughout life, with each new set being slightly larger than the previous one. In addition to increased size, the number of teeth may also increase with age. The concept of continuous replacement can be confirmed by looking at any adult animal, where there will always be one or two sets of replacing teeth present

Nothing but the Tooth. DOI: http://dx.doi.org/10.1016/B978-0-12-397190-6.00009-2

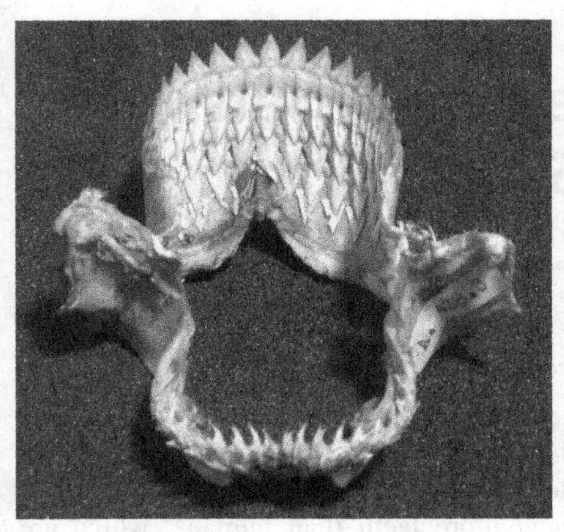

Figure 9.1 Jaw of kitefin shark (Dalatias family). The teeth in the upper half of the image are from the lower jaw and are blade-like and locked together. Beneath the functioning set, four or five replacing sets of teeth are evident. In life there would have been more but, as these are less well developed, they are lost during preparation of the specimen. The teeth in the lower part of the image are from the upper jaw and are more conical in shape.
Source: Courtesy of the Hunterian Museum at the Royal College of Surgeons. Photographed by M. Farrell.

underneath the gums regardless of age (see Figure 4.15). In humans, no replacing teeth will be present after about the age of 12 years.

The easiest way of demonstrating the presence of numerous sets of teeth is to look at the jaws of a shark. The shark is exceptional in that in addition to one or more sets of teeth in function in the mouth, up to 10 further sets of replacing teeth at successively younger stages of development may be present beneath the lining of the mouth (Figure 9.1).

In young sharks, teeth are replaced every 2–4 weeks. The belief that tooth replacement occurs to compensate for the wear and blunting of teeth is incorrect, as little or no tooth wear occurs during the relatively short time the teeth are in function. To confirm this assumption, if a shark is deprived of food, tooth replacement still occurs at the normal rate.

It is possible to estimate the total number of tooth replacements throughout life by measuring the width of a tooth and that of the slightly larger replacing one beneath it in a series of skulls ranging from very young to very old and relating it to jaw growth using a mathematical formula (Strasborg plot). In the piranha (see Chapter 1, page 8–9) nearly 30 sets of teeth are produced during its lifetime. In the Nile crocodile, up to 50 replacements occur at each tooth position during its long life.

Tooth replacement patterns have been studied in living nurse sharks on a weekly basis for 3 years. The functional teeth were marked regularly to identify them. It was seen that rows of teeth were replaced every 10–20 days in the warmer summer months, but slowed to 50–70 days in the colder winter months. If these rates are extrapolated over the life of the animal, the number of tooth replacements is far larger than the 50 sets reported for the Nile crocodile.

In the case of sharks, therefore, literally thousands of teeth are produced over a 10–15-year lifespan, yet most show little sign of wear before being replaced. This may seem a very wasteful and inefficient system, especially when considering the largest known fossil shark, Megalodon, whose massive teeth could reach a length of 15 cm in (Figures 9.2 and 9.3). However, it must be kept in mind that sharks are

Figure 9.2 Photograph of Megalodon from the American Museum of Natural History. This early reconstruction is now agreed to have made Megalodon considerably larger than it really was. Man sitting on *Carcharodon megalodon* jaws. Pre-1923.
Source: Posted by Funkmunk, 2009. http://en.wikipedia.org/wiki/File:Carcharodon_megalodon.jpg#filelinks

Figure 9.3 Two teeth of Megalodon with a human molar tooth for comparison.
Source: Courtesy of the Hunterian Museum at the Royal College of Surgeons.

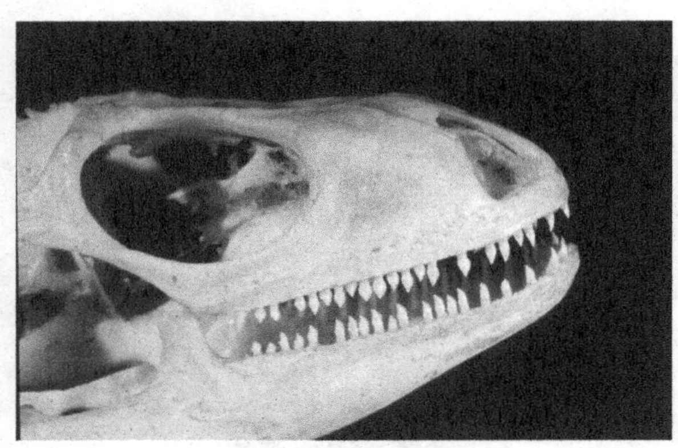

Figure 9.4 Skull of iguana. The teeth all have the same somewhat conical shape. At any one time, it is common to have a few teeth missing, with gaps that will be replaced by newly erupted teeth. Note, however, that in this skull no two adjacent teeth are missing.

a hugely successful species, having been around for hundreds of millions of years in more or less the same basic form, so it can't be that much of a handicap to a shark to replace them so rapidly. Some sharks have been observed to ingest exfoliated teeth, perhaps nature's way of reutilising the calcium and phosphates in them.

As the teeth of non-mammalian vertebrates enlarge with age, they are retained for progressively longer periods, so that the rate of tooth replacement slows down with age. Whereas in a young piranha, lizard or crocodile, a tooth may function in the mouth for a period of weeks, in older individuals it lasts for months.

Although in sharks many adjacent teeth are shed at the same time, the underlying replacing teeth erupt very quickly so that there is little interruption to feeding. In the majority of non-mammalian vertebrates, as typified by lizards, there may be a delay of a few weeks between the shedding of one tooth and its replacement by another. At any one time gaps will be present in the jaws, as typified by a lizard such as an iguana (Figure 9.4). If a number of adjacent teeth were shed at the same time, then that part of the jaw may be inefficient during food gathering. However, this situation is rarely encountered, as the majority of teeth are always present in the mouth with just a few isolated gaps here and there, beneath which replacing teeth are preparing to erupt. The question arises as to whether this observation is the result of chance or whether there is some additional mechanism involved.

If you are good at solving puzzles, then look at Figure 9.5. It is a model of a real lizard jaw showing the teeth that have erupted as well as the replacing teeth lying beneath. Only three gaps are evident in the 24 tooth positions, namely positions 7, 14 and 19, where teeth have dropped out but have not yet been replaced. The replacing teeth within this jaw are at varying stages of development; some are very small, having just started to develop (tooth positions 6, 8, 15), while others are larger and therefore older and more advanced, and will soon erupt (tooth positions 7, 14, 19 and 22). Can you identify any orderly pattern in which the replacing teeth will erupt?

There does not immediately appear to be any obvious pattern whereby the functional teeth are replaced when looking at the model in terms of adjacent teeth. However, if alternate teeth are examined, a pattern does emerge (Figure 9.6).

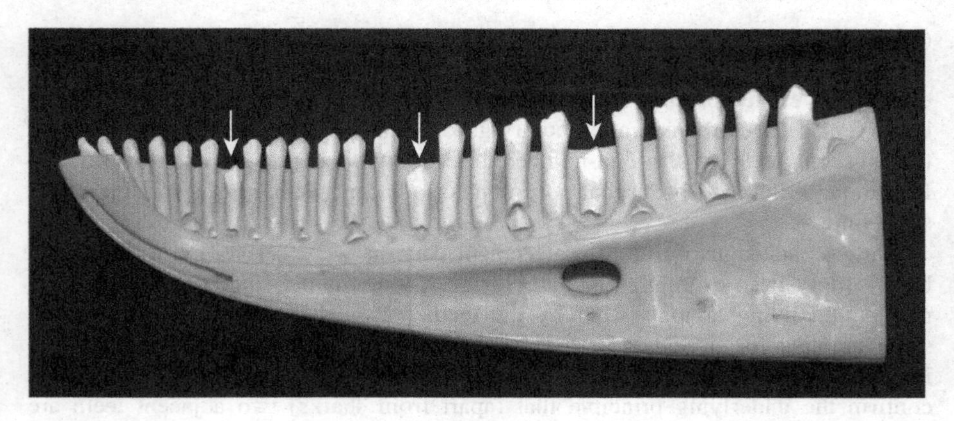

Figure 9.5 Lower jaw of green lizard. Notice the gaps (arrows) where teeth have dropped out and the replacing teeth underneath will soon erupt. Also notice that beneath each erupted functional tooth are small, developing replacing teeth at varying stages of development.
Source: Courtesy of Dr J.S. Cooper.

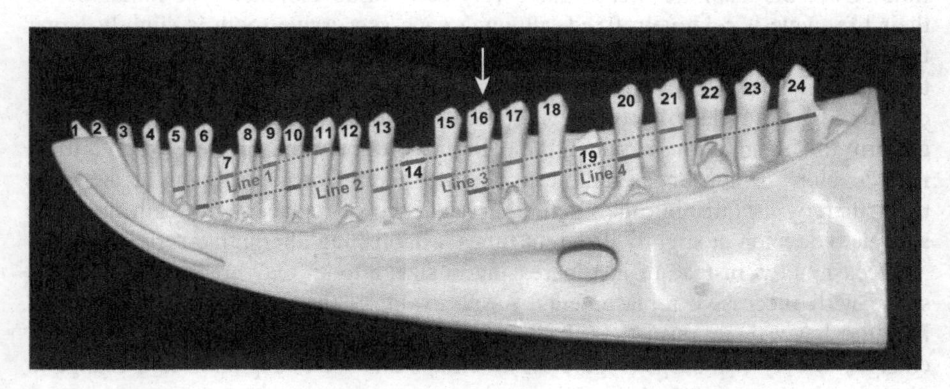

Figure 9.6 Same model as in Figure 9.5. The arrow indicates tooth position 16. Lines have been drawn through progressively younger replacing teeth in alternate tooth positions, indicating the order in which they will erupt. Lines 1 and 3 join odd-numbered teeth and lines 2 and 4 even-numbered teeth. See text.
Source: Courtesy of Dr J.S. Cooper.

Focusing first on, say, tooth position 16 (arrow), this tooth has just erupted and there is no sign yet of a replacing tooth beneath it. Moving forwards through tooth positions 14, 12, 10, 8 and 6, the replacing teeth are at progressively younger stages of development. It can be assumed, therefore, that these teeth will erupt from back to front in this alternate sequence (Figure 9.6, line 2). Similar alternate ways of replacement will affect all the teeth, some passing through the even-numbered teeth (Figure 9.6, lines 2 and 4), others passing through the odd-numbered teeth

(Figure 9.6, lines 1 and 3). Depending on the speed of replacement, only one tooth in any wave is likely to be missing at any one time. As the waves affecting the odd- and even-numbered tooth positions are out of synchrony, whenever a tooth is being replaced and there is a temporary gap, let us say in an even-numbered tooth, the adjacent teeth in the odd-numbered series will be present and functioning. In summary, this will mean that no groups of adjacent teeth will be missing at the same time.

These conclusions on tooth replacement patterns were derived on a theoretical basis simply from examining skulls. However, longitudinal studies have been carried out in living animals whereby the teeth present have been recorded in the same animals for periods of over a year using techniques such as taking wax impressions (for trout and lizards) or radiographs (for alligators). These studies confirm the underlying principle that (apart from sharks) two adjacent teeth are rarely shed at the same time, but that a series of waves of replacements pass through alternate tooth positions, with those in odd- and even-numbered positions being 'out of sync' with each other.

What is the mechanism responsible for ensuring that teeth are replaced in this alternating pattern and that groups of adjacent teeth are not all shed at the same time? Once the teeth are first initiated very early in development, the remainder of their life cycle is relatively fixed, although each successive tooth is slightly larger than the preceding one and generally lasts for a longer period. As long as adjacent teeth are first initiated at different times, they will not be shed together. Studies determining the order of tooth development in the trout, frog, lizard and alligator confirm this assumption, although the pattern varies between species. When a tooth first develops, it appears to inhibit an adjacent tooth from developing at the same time, thereby establishing the alternating sequence of replacement: teeth therefore appear to develop at slightly different times, establishing alternating series.

Occasionally, instead of just increasing in size, a tooth may gradually change its shape with successive replacements. As an example, the teeth of the green lizard initially have three cusps. With successive tooth replacements, this gradually changes to only two cusps (Figure 9.7). Perhaps this is an adaptation to a change in diet or a change in eating behaviour.

Within the vast animal kingdom, there will always be exceptions to the general rule. For example, the chameleon has only a single set of teeth that are slowly worn down with age to the underlying bone. This bony ridge then continues to act as an efficient cutting/grinding surface. The reader is referred to the piranha (Chapter 1) for a unique situation where teeth develop and *are* replaced simultaneously.

Tooth Replacement in Cichlids

One of the most remarkable, and as yet unexplained, specialisations of tooth replacement occurs in fish of the cichlid family. Commonly known as Alluaud's Haplo, this African cichlid is found in East African lakes and rivers of the Lake Victoria drainage system. It possesses teeth not only on its jaws but also on bony structures located

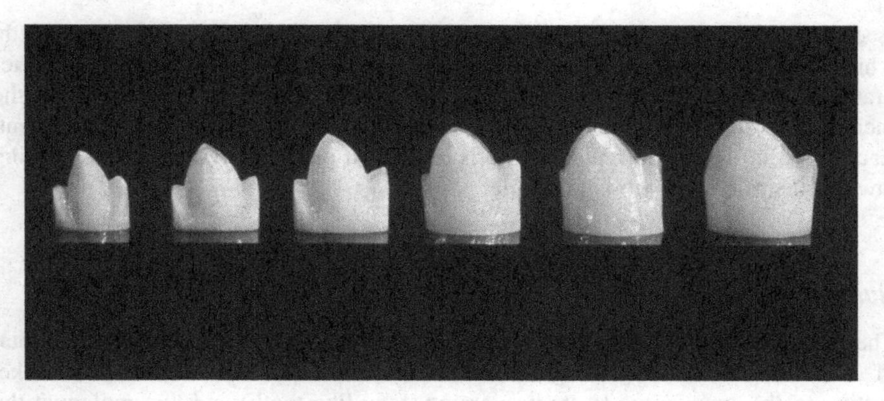

Figure 9.7 A model of a tooth of the green lizard showing the gradual change in morphology from a tooth with three cusps in a young specimen on the left to one with two cusps in an adult tooth on the right following a number of tooth replacements.
Source: Courtesy Dr J.S. Cooper.

farther back in its throat (pharynx). Rather than carrying gills, these structures serve as a secondary set of jaws. These pharyngeal teeth help to process the food. Like the teeth in the mouth, they also show typical continuous tooth replacement. When eating soft food, these fish develop numerous sharp, pharyngeal teeth, but when feeding on a durophagous diet (e.g. snails with hard shells), they have fewer but broader pharyngeal teeth. Experiments have shown that if the diet is changed from hard to soft, the form of the pharyngeal teeth also changes. Thus, young durophagous cichlids, when fed a soft diet instead of their normal snails, will develop numerous sharp, pointed teeth instead of the expected fewer but broader teeth. How the change in the environment works on the genes to affect a change in the shape of the teeth remains a mystery. It is the equivalent of feeding the fearsome *Tyrannosaurus rex* a vegetarian diet and expecting its teeth to become flat!

Newly Hatched Alligators

In the embryos of many reptiles, rudimentary teeth start to develop but are lost early without ever becoming functional. Newly hatched alligators are small versions of the adult, already possessing a functioning set of small erupted teeth, which allows them to feed independently. This functioning set, however, is not the first to develop, as four sets of teeth have already developed and disappeared before the alligator hatches.

Tooth Replacement in Humans and Other Mammals

In addition to crowns, the teeth in mammals differ from those of non-mammalian vertebrates in having roots that are supported in bony sockets in the jaws (see Figures 4.1 and 4.2). They also differ in possessing only two sets of teeth, which have to last throughout their whole lives. Unlike non-mammalian vertebrates, mammalian growth stops at maturity. The teeth accommodate to the increased jaw size up to maturity without the necessity of continuous replacement. Each tooth in the single replacing set is integrated periodically within the growing jaw when sufficient room is available.

Humans

The first set of milk or deciduous (shed) teeth number 10 in each jaw, giving a total of 20. On each side, beginning from the front, there are two incisors (blade-like, cutting teeth), one canine (a stout, piercing, fang-like tooth) and two molars at the rear (broader teeth with cusps for grinding up food). The milk teeth erupt at periodic intervals as space becomes available in the growing jaws. The incisor teeth are the first to appear at about 6 months of age, while the last are the back teeth (second molars) at about two-and-a-half years. For the next 3 or 4 years, the infant has a complete milk dentition of 20 teeth (Figure 9.8).

Occasionally, the milk incisors erupt early and are present at birth; these are referred to as natal teeth. Many superstitions have centred around such infants. In some affected newborn infants, such as Hannibal, Louis XIV and Napoleon Bonaparte, the presence of natal teeth was regarded as a highly favourable sign. In others, such as Richard III and Ivan the Terrible, it was considered a sign of evil. Children born with teeth were often the subject of infanticide (which no doubt continues in some parts of the world) as it was associated with witchcraft. Having an extra finger, as did Ivan the Terrible, would not have helped their cause either.

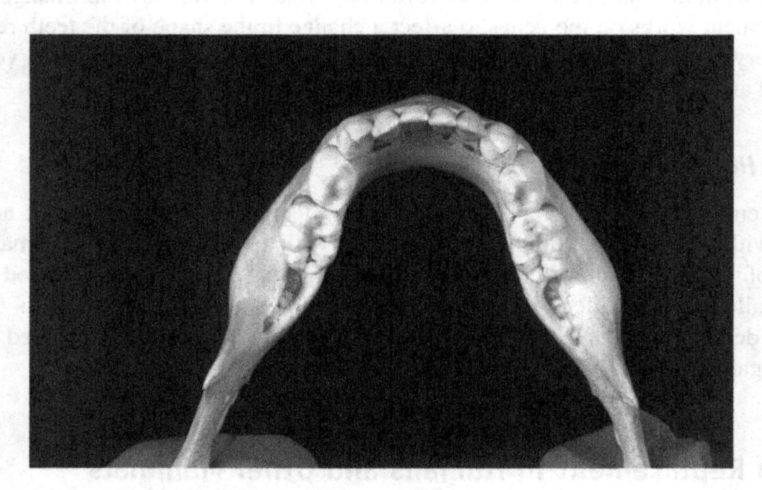

Figure 9.8 Lower jaw of a child aged 3 years, showing five milk teeth on each side. The back teeth are the second deciduous molars and those by the midline in the front are the first (central) incisors.

DISEASES and CASUALTIES.

Disease	Number	Disease	Number	Disease	Number
AGED	1563	Gangrene	16	Rupture	23
Ague	7	Gout	26	St Anthony's Fire	10
Apoplexy	71	Green Sickneſs	1	Scald Head	1
Aſthma	3	Grief	6	Scarlet Fever	7
Bedridden	3	Griping in the Guts	985	Scurvy	14
Black Catterick	1	Headmouldſhot	9	Small Pox	898
Bleeding	4	Hectic Fever	1	Sores and Ulcers	45
Bloody Flux	6	Hiccough	1	Spleen	3
Burſten	3	Hooping Cough	5	Spotted Fever	74
Cancer	77	Jaundies	102	Stone	39
Canker	9	Impoſthume	49	Stoppage in the Stomach	333
Childbed	217	Infants	25	Strangury	9
Chriſomes	51	Lethargy	5	Strongullion	3
Colick	102	Livergrown	6	Suddenly	68
Conſumption	2831	Lunatick	34	Surfeit	70
Convulſion	5493	Meaſles	51	Teeth	1305
Cough	3	Mortification	14	Thruſh	41
Cut of the Stone	1	Pain in the Head	1	Tiſſick	310
Diabetes	4	Palſy	20	Twiſting of the Guts	1
Diſtracted	2	Pleuriſy	23	Tympany	8
Dropſy	848	Purples	11	Vomiting	9
Evil	101	Quinſy	14	Water in the Head	11
Fever	3162	Raſh	5	Wen	1
Fiſtula	27	Rheumatiſm	26	Wind	1
Flux	5	Rickets	381	Wolf	1
French Pox	63	Riſing of the Lights	86	Worms	47

CASUALTIES.	
Abortive	111
Bruiſed	2
Burnt	3
Choaked	2
Dead by Misfortune (ſo reported)	1
Drowned	61
Executed	6
Found dead in the Streets, &c.	19
Frighted	1
Hang'd and made away themſelves	21
Killed by ſeveral Accidents	72
Murdered	12
Overlaid	69
Scalded	1
Stifled	3
Stillborn	443
Suffocated with Charcoal	1

CHRISTENED		BURIED	
Males	7765	Males	10354
Females	7683	Females	10366
In all	15448	In all	20720

Of the Plague 0

Increaſed in the Burials this Year 1239

Figure 9.9 Page from Bills of Mortality for the parishes of London for the year 1703.
Source: Courtesy of the Library of the Royal Society of Medicine, London.

During the first 2 or 3 years, as the milk teeth erupt, the child may experience some discomfort. The symptoms of this 'teething' will vary. The gums may be sore, the cheeks red and the child irritable and with a fever. Up to the eighteenth and nineteenth centuries, teething was mistakenly believed to be a direct cause of death in young children. Figure 9.9 shows the recorded deaths in London for the year 1703. Of a total of 26,720 deaths, 1305 are listed as being due to 'teeth'. Of course, the real causes had nothing to do with teeth and were all about the unhygienic and squalid living conditions, the lack of suitable food and infectious diseases, especially in the poor.

As the child grows, the jaws enlarge, as is evidenced by spaces appearing between the teeth. A larger tooth surface is now needed to deal with the increased and more varied food ingested by the growing child. This is achieved in two ways, namely by providing replacement teeth that are larger and by having more teeth. Figure 9.10 is the skull of a child aged five and a half but with bone removed to show the underlying permanent teeth at varying stages of development and the first permanent molars about to erupt behind the back deciduous tooth.

In humans, tooth replacement takes place over a period of about 12 years, from the ages of 6–18. Each of the first three front milk teeth is replaced by a permanent tooth that is larger. (The back two milk teeth, the deciduous molars, are replaced by two permanent premolars that are actually very slightly smaller in

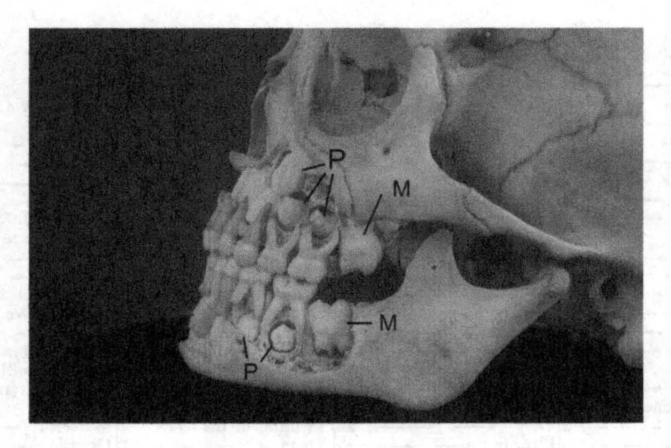

Figure 9.10 Left side of the skull of a five-and-a-half-year-old child showing the erupted five deciduous teeth in each jaw. The bone has been dissected away to show some of the permanent teeth (P) developing beneath the erupted deciduous teeth and the first permanent molar (M) about to erupt behind the last deciduous tooth.
Source: Courtesy of the Hunterian Museum at the Royal College of Surgeons.

length.) In shedding, the roots of milk teeth are eroded/resorbed by the underlying erupting permanent tooth, resulting in their looseness. If a permanent tooth fails to develop below its milk tooth, the milk tooth may be retained into adult life.

The milk teeth are shed and replaced periodically by permanent teeth, with the permanent lower first incisors erupting first, at about 6–7 years, and the second permanent premolars (or bicuspids, meaning two cusps) erupting last at about 11–12 years. From about the ages of 6–12 years, the erupted dentition is composed of a mixture of milk and permanent teeth, the 'mixed dentition' stage (Figure 9.11). Following its initial appearance, a permanent tooth takes about a year to reach the required height to bite against its antagonist tooth in the opposite jaw. Although gaps appear periodically following the shedding of a milk tooth and before the eruption of the permanent successor, the gaps do not prevent the child from eating properly.

As each jaw increases in length, sufficient room becomes available at the back for the addition of three large grinding teeth on each side, the permanent molars. These teeth lack predecessors. These molars are added periodically about every 6 years: the first permanent molar teeth erupt at 6 years of age behind the last milk tooth (see Figure 9.10), the second permanent molar teeth behind it at about 12 years of age (Figure 9.11) and the third permanent molar teeth (the so-called 'wisdom' teeth because one is older and wiser by the time they erupt) at about 18 years of age. Each jaw eventually has 16 teeth (eight on each side) (Figure 9.12), giving a combined total of 32. The complete adult dentition starting from the front has two incisors, one canine, two premolars (replacing the two deciduous molars) and three molars on each side.

When the permanent molar teeth first erupt with their smooth, unworn rounded cusps, they are not well suited to chewing and require a good period of 'wearing

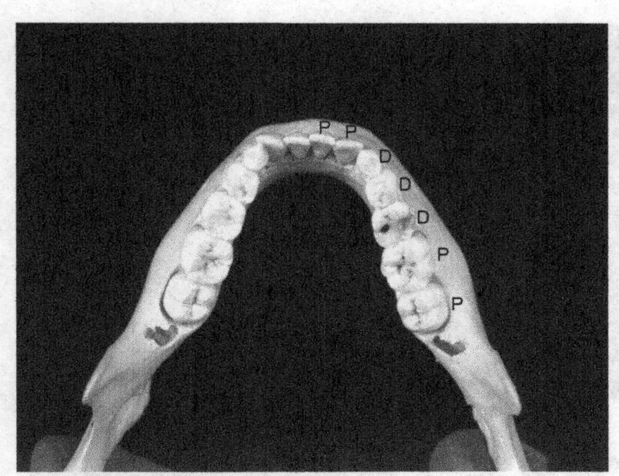

Figure 9.11 Lower jaw of a child aged 11 years, showing a mixed dentition, with both deciduous (D) and permanent (P) teeth. The back tooth represents the second permanent molar that is just erupting into the mouth.

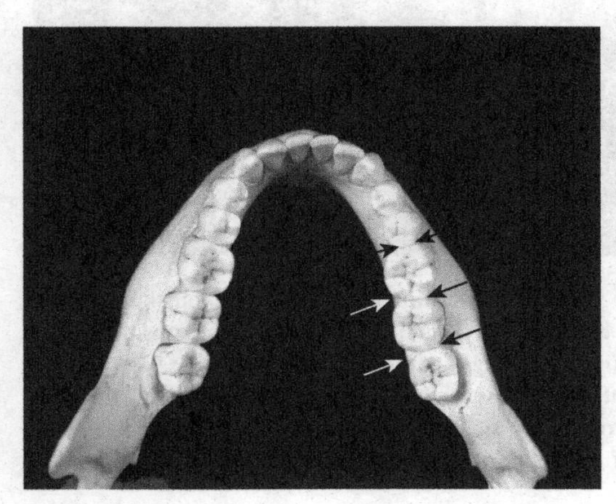

Figure 9.12 Lower jaw showing the complete adult dentition consisting of eight teeth in each half. As it is a young adult of about 25 years of age, the teeth show little wear and the contact points between adjacent teeth are small (arrows). Compare with Figure 9.15.

in' to flatten them a little for better function. Can you imagine how inefficient it would be if, just when they were finally worn in and functioning well, the permanent molars were suddenly replaced and the whole process had to be started all over again! (This is further illustrated in Figure 9.16.)

In humans today, the wisdom teeth are absent in up to 25% of the population, or, if present in the lower jaw, they remain unerupted (impacted) in a further 25% of the population due to lack of space (Figure 9.13). If partially erupted, food stagnation can occur around an impacted molar and result in painful inflammation of the gums. This is the main reason for the extraction of wisdom teeth.

One explanation for the widespread occurrence of impacted wisdom teeth concerns the nature of the modern diet in many so-called 'advanced' countries. In 'primitive' societies, the diet is often tough, coarse and abrasive, requiring considerable chewing. This results in rapid wear of much of the grinding surface. The wear may be so extensive as to expose the dental pulp and cause painful abscesses. This is typically seen in skulls from early civilisations (Figure 9.14) where, for example, the bread contained quantities of grit and other abrasive

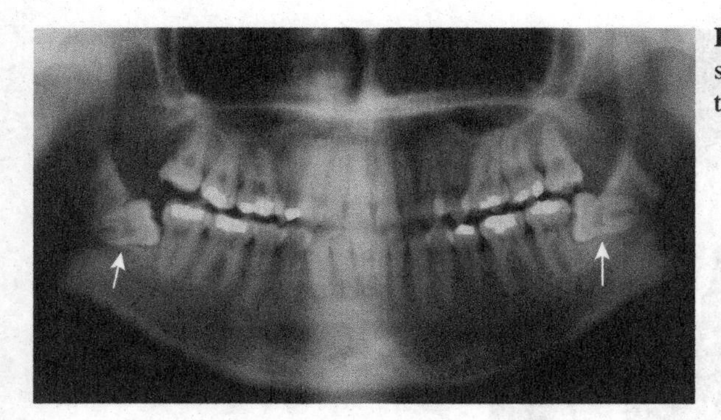

Figure 9.13 X-ray of dentition showing two impacted lower third molars (arrows).

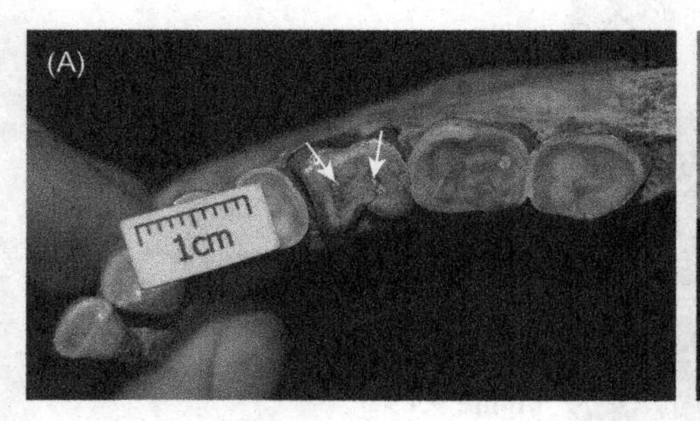

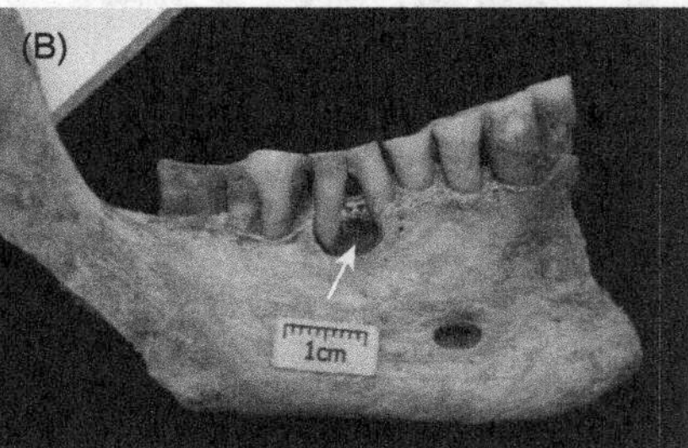

Figure 9.14 (A) Viewed from above, a well-worn dentition of an Anglo-Saxon skull 1200 years old showing very severe wear on all the teeth. This has resulted in the exposure of the dental pulp of the first molar tooth (arrows), causing a painful abscess. (B) Side view of same skull as in (A), showing the site of dental abscess (arrow) beneath the roots of the first molar tooth. Note the severe flattening of the top surfaces of the crowns.
Source: ©Melanie Nichols. Courtesy of the Hunterian Museum at the Royal College of Surgeons

particles during its preparation. As teeth move very slightly in their sockets during chewing, there is also wear at the contact points between adjacent teeth: these contacts are narrow when the teeth first erupt due to their natural bulbous outline. This wear might be expected to leave small gaps between the teeth. However, by some unknown mechanism, these gaps automatically close up as the teeth all move slowly forwards (mesial drift). Eventually, with further wear, the contact points between adjacent teeth become much broader (compare Figure 9.12 with Figure 9.15). In groups such as Eskimos and Australian Aborigines, whose diet (at least until recently) consisted of tough food, this produced as much as 3–4 mm of wear at the contact points on each side of the jaw. As the teeth move forwards to fill the gaps, this allows ample room at the back of the jaw for the wisdom teeth to erupt, so impacted teeth hardly ever occur in these populations.

The soft, processed and cooked foods eaten by the majority of present-day North Americans and Europeans mean that not a lot of vigorous chewing takes

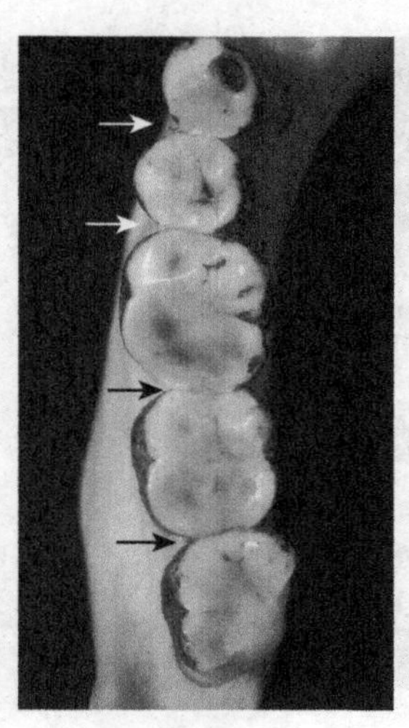

Figure 9.15 Well-worn dentition showing the loss of tooth substance between adjacent teeth, resulting in a very broad contact point (arrows). Compare this with the narrow contact point in Figure 9.12. As the teeth move forward to close up the potential space (mesial drift), enough room is provided at the back of the jaw to allow the last molar tooth to erupt normally and not become impacted.

place in the population. The teeth are not worn as much and hence they do not move forwards much. This results in a lack of space at the back of the jaw to accommodate the erupting wisdom teeth, which frequently become impacted (Figure 9.13), providing a never-ending supply of patients for the oral surgeon. Perhaps if we all chewed on seal skin for half an hour every day, impacted wisdom teeth would be a thing of the past!

In humans (and mammals) the permanent set must last at least until reproductive age has been reached and ideally for some time beyond. With advances in medicine, humans today can easily live up to 80 years and more. Features that have evolved to make mammalian teeth last longer include their large size (particularly in some of the bigger herbivores, such as the horse and cow; see Chapter 4) and the presence of a very thick layer of enamel, the hardest and most resistant substance in nature. In humans, the enamel layer can have a maximum thickness of 2.5 mm, whereas the equivalent layer covering the dentine in non-mammalian vertebrates is normally very thin (about 0.2 mm) and with a simpler structure. The presence of roots attached to the bony socket by a fibrous membrane allows for slight movements to maintain tooth contact after tooth substance has been worn away.

Other Mammals

Crushing/cutting the food and absorbing its energy quickly is important in warm-blooded mammals. In many animals, the initial form of the teeth when they first erupt is not ideally suited to their grinding function. Take for example the grinding molar teeth of the hippopotamus. When they first erupt, the crowns possess smooth, rounded cusps that are inefficient in grinding up the tough vegetable matter on which the animal feeds (Figure 9.16A). After a 'wearing-in' period, the cusps are flattened to expose a complex pattern of folded ridges of enamel. Between the white enamel ridges are brownish areas of softer dentine and cement. As these

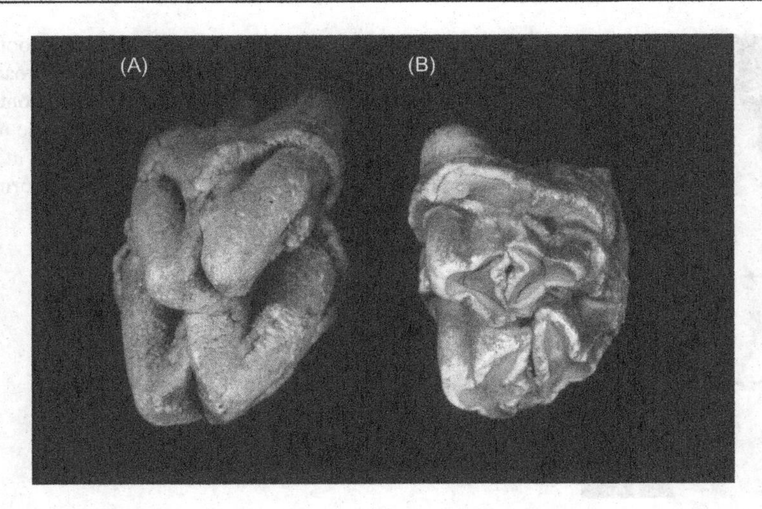

Figure 9.16 (A) View of unworn hippopotamus molar tooth with smooth surface. (B) Worn hippopotamus molar tooth. Note the roughened surface, with the white enamel ridges standing 'proud' of the brownish dentine and cement to provide a suitable roughened surface suited to masticate the food.

three exposed dental tissues have different hardness levels, they wear away at different rates. Consequently, chewing food will leave the harder, more resistant, enamel at a higher level than the slightly softer dentine, ensuring that a roughened surface ideally adapted to masticate tough, abrasive vegetation is automatically maintained (Figure 9.16B). Imagine if the teeth, having been worn in and now suitable for function, were suddenly replaced by a new set that again required a lengthy 'wearing-in' process. This would clearly reduce the masticatory efficiency of the teeth. Hence, only two dentitions are present during life.

In nature, there are always exceptions to any rule. There are some mammals that only have a single set of teeth, such as toothed whales (dolphins, porpoises, sperm and killer whales) and many rodents (e.g. rats and mice). In the toothed whales, the teeth all have a simple cone shape and are used for grasping their prey, usually other fish and squids, prior to swallowing it whole, and not for grinding the food.

Although rats and mice have only a single dentition and only three cheek teeth, their enormous success as a species is partly due to the specialisation of their front incisor teeth, always prominent in animated cartoons (such as *Bugs Bunny*). These incisor teeth have the remarkable property of continuous growth, enabling them to last throughout life. The biting edges of the teeth worn away by gnawing are replaced by newly formed tooth material produced at the base, which lies deep within the bony socket of the jaw (Figure 9.17). The incisors continually erupt in their sockets, although their length remains the same, so the whole effect is like that of a moving escalator.

Rats normally gnaw away about 3 mm of tooth substance from their tips every week, but this is compensated for by renewing the same amount of new tissue

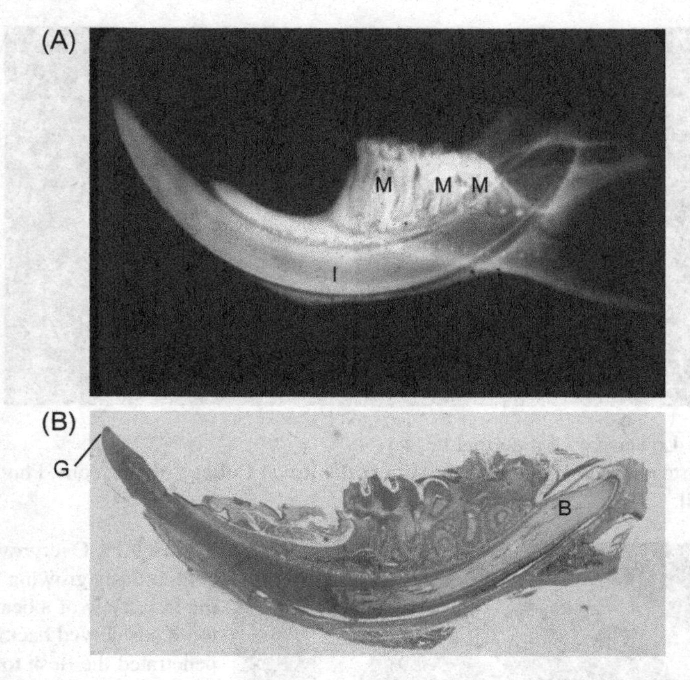

Figure 9.17 (A) X-ray of the lower jaw of a rat, showing the large continuously growing incisor (I) that occupies most of the jaw. There are only three other teeth in the jaw, the molars (M). (B) Microscope section of the lower jaw of the rat. The tissues at the base of the incisor (B) continually provide new dental tissue throughout life to compensate for loss occurring at the gnawing edge (G).

at their bases. The incisor tooth has the potential of raising the rate of production to 7 mm a week. This provides the animal with a wide safety margin if, for example, the tooth should fracture or the food is very tough and abrasive, requiring extra wear. Although having continuously growing teeth might seem the answer to all our prayers in the case of humans, consider what happens in animals should the opposing teeth become misaligned and unable to meet. In such situations, the teeth will continue to grow and erupt past each other. They may then become a physical obstacle to chewing and result in the death of the animal. Incisors from a normal beaver are shown in Figure 9.18. The lower incisor of a beaver that has erupted beyond its corresponding upper tooth is shown in Figure 9.19. It has continued to grow in a circle and has penetrated the skin and come to lie inside the back of the lower jaw. Life for this particular beaver couldn't have been much fun. People with pets such as hamsters and rabbits often have to visit their veterinary surgeon to have the teeth clipped in order to control this problem of overgrowth (Figure 9.20).

An animal's size can have a significant effect on tooth replacement. Shrews, though successful mammals, are short-lived. They have a very rudimentary set of milk teeth, and the permanent teeth commence erupting within a few days of birth.

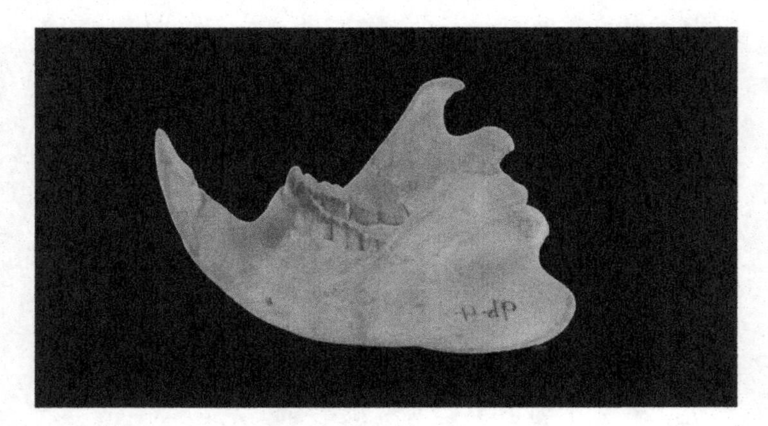

Figure 9.18 Lower jaw of a normal beaver.
Source: Courtesy of the Hunterian Museum at the Royal College of Surgeons. Photographed by M. Farrell.

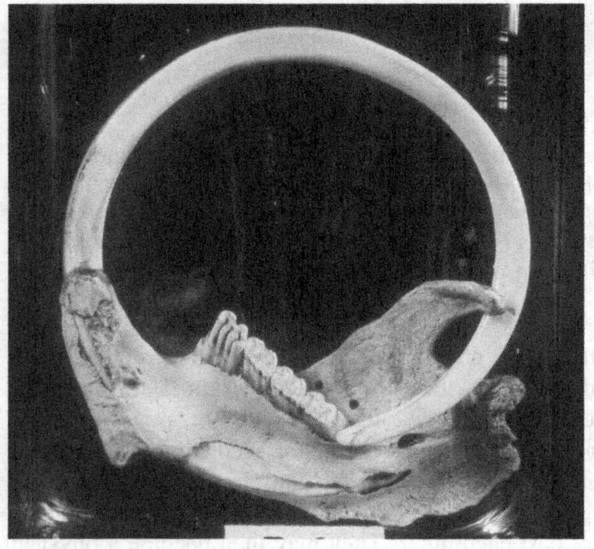

Figure 9.19 Overgrowth of continuously growing incisor in the lower jaw of a beaver. The tooth has curved backwards and penetrated the flesh to contact the bone of the jaw.
Source: Courtesy of the Hunterian Museum at the Royal College of Surgeons.

After about 18 months, the teeth have worn away and the animal dies. At the opposite end of the spectrum where size is concerned is the long-lived elephant. Its food is extremely abrasive, and it requires very large teeth to get through it. It is not possible to accommodate six pairs of large teeth on each side all at the same time, even if it could bear the weight of them. The solution that has evolved is that the elephant only ever has one pair functioning at any one time, their size gradually increasing with each replacement. The last molar tooth can weigh over 10 kg! Once the first pair has worn down after a few years, the next pair erupt from behind and displace them (see Figure 2.4). When the sixth and last pair on each side has been worn out in very old elephants, the animal will die.

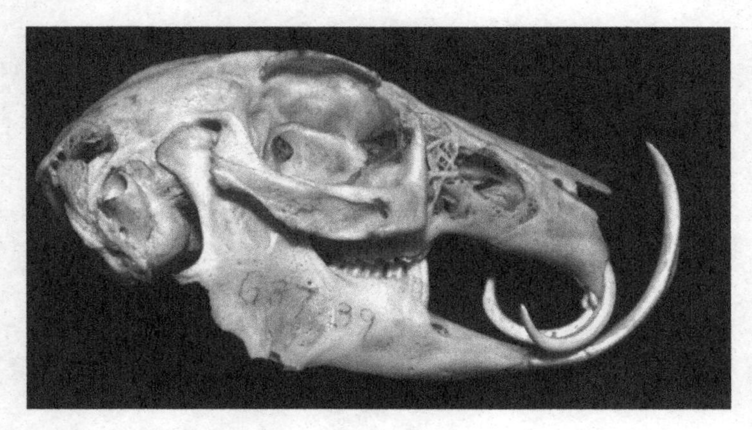

Figure 9.20 Overgrowth of rabbit incisors.
Source: Courtesy of the Hunterian Museum at the Royal College of Surgeons.

Summary

Although it sounds like a good idea, there are drawbacks in having continuous tooth replacement like sharks and other non-mammalian vertebrates, which grow throughout life. Mammals do not constantly have to adapt to an increasing size of jaw because they cease to grow at maturity. Continuous tooth replacement is particularly inefficient in mammals where teeth bite together to masticate food. They have evolved different strategies to cope with having only two sets of teeth during their lives. In humans, the ultimate solution is to get a third set from the dentist!

10 Serendipity and the Discovery of the Modern Dental Implant

Important scientific discoveries are sometimes the result of an accident or chance (serendipity). A scientist may be involved in experiments on one topic when an unrelated observation appears. In some cases this observation is ignored; in others it may trigger thoughts worthy of further consideration, leading the scientist into a completely different field of study.

One of the most famous examples of serendipity concerned an observation that led to the discovery of the antibiotic, penicillin. Dr Alexander Fleming (later Sir) was a bacteriologist who had already discovered an enzyme in tears, called lysozyme, which was able to kill bacteria. In 1928, he was studying a group of bacteria, the *Staphylococci*, well known for causing disease. Not being renowned for his cleanliness in the laboratory, Fleming left for a few days without tidying up. On his return he came across some old culture dishes without their lids that had been lying around exposed to the atmosphere. He noticed that one dish was partly contaminated with a mould. The unaffected part without any mould still contained bacterial colonies of living *Staphylococci*. However, the zone adjacent to the mould was completely devoid of bacteria. Fleming realised that the mould must have released a substance that killed the bacteria in its immediate vicinity.

Following up this chance observation, Fleming identified the mould as belonging to the penicillin family of bacteria, and he named the bacteria-killing substance 'penicillin'. He further discovered that penicillin was lethal to a number of other bacteria responsible for serious illnesses, such as pneumonia, scarlet fever and meningitis, and in 1929 he published an account of his findings. However, working on his own and with limited resources, Fleming found the penicillin mould difficult to culture and the bactericidal agent, penicillin, hard to extract. He also suspected that penicillin might not remain active once injected into humans, so he abandoned the research.

In the late 1930s, Drs Howard Florey and Ernst Chain (both later knighted) were working in England on bactericidal agents and rediscovered Fleming's earlier report on penicillin. With greater determination, belief and expertise, they went on to isolate and produce penicillin in sufficient quantities to inject into a few patients. Those early trials demonstrated penicillin's enormous potential as an antibacterial agent. The question then arose as to whether the production process of penicillin should be patented. This could engender enormous financial rewards. Although there was some disagreement among the research team, the decision was made by

Nothing but the Tooth. DOI: http://dx.doi.org/10.1016/B978-0-12-397190-6.00010-9

Florey not to patent it, but to give it freely to the world. (A similar moral problem arose with the discovery of general anaesthesia; see Chapter 3.)

With the onset of the Second World War, production of penicillin was moved to North America, the only country where it could be produced in the large, industrial quantities required, although this meant that the financial profits would go to America. In 1953, Florey, Chain and Fleming were awarded the Nobel Prize for Medicine for their discovery of penicillin.

Chapter 6 describes how serendipity played a role not once, but twice in the discovery of growth factors, resulting in the award of another Nobel Prize in Medicine. This chapter describes how serendipity was involved in one of the most important discoveries associated with dentistry in the last 60 years, the modern dental implant.

In humans, the roots of teeth are attached to the adjacent bone by a fibrous joint (periodontal ligament; Figure 10.1A). As humans have only one set of adult teeth (see Chapter 4), a main goal in dentistry is to replace teeth when they are lost by disease or trauma. The ideal solution might be to replace a missing tooth with one grown in a test tube using modern tissue-engineering techniques. For this to succeed, the new tooth must be accepted by the body as 'self' and not be rejected as 'foreign'. Even if this technique eventually becomes available, it may well prove too lengthy and expensive for general use. Until recently, two viable solutions were on offer: a removable denture or a fixed dental bridge.

A removable plastic denture may be either partial (if only a few teeth are replaced) or full (if all the teeth are replaced). A full denture may lack sufficient support from the underlying bone, rendering it loose and causing difficulty in eating and speaking. Even partial dentures are not always easy to fit and may require additional clips on adjacent teeth to provide stability. As dentures can cover a

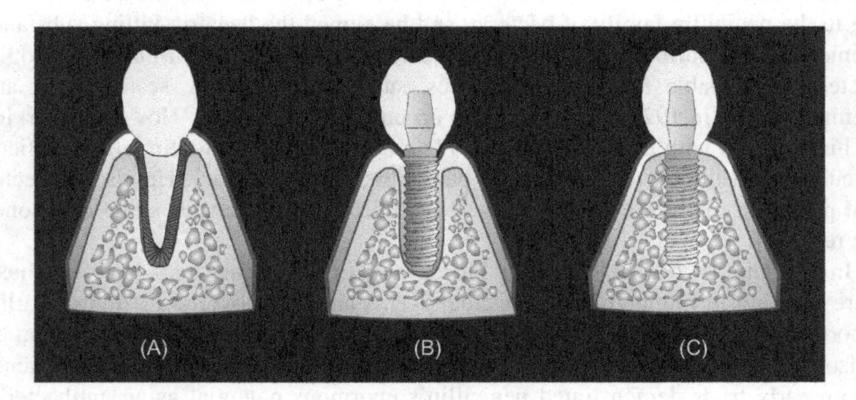

Figure 10.1 (A) A normal tooth attached to the bone by a special fibrous membrane called the periodontal ligament. (B) A non-titanium implant separated from the bone by a fibrous membrane. (C) A titanium implant directly bonded to bone (osseointegration), without the presence of an intervening fibrous layer.

considerable area of the mouth, they may reduce the sensation of taste and perhaps look 'false'.

A fixed bridge, on the other hand, is more stable but has the disadvantage of necessitating the cutting away of sound tissue in adjacent teeth, which are then capped to support a false tooth that fills the space (Figure 10.2).

A third solution to replacing a tooth or teeth would be to use what is termed an 'implant'. This involves inserting a pin (the implant) into the bone of the jaw where a tooth is missing that acts as a support (root) on which an artificial tooth can be cemented. There has been a long history of attempts to find a material suitable for inserting into the jawbone that would remain firm and not be rejected. These materials have included gold, steel, ivory and even sea shells and coral. Made from any of these materials, a fibrous membrane always appeared between the implant and bone, without a firm attachment to either (Figure 10.1B). This resulted in continuous slight movements of the implant, eventually causing it to become loose and necessitating its removal. There was also a tendency for inflammation to occur both in the fibrous membrane and the site where the implant penetrated the gum.

The eventual solution of finding a stable implant material arose purely by chance. The discovery was not made in one of the countries with the greatest number of dental schools and research institutes, such as the United States or the United Kingdom, but in Sweden. Furthermore, the discoverer had no connection with dentistry, being neither a dentist nor a dental scientist.

The story of the dental implant began in the University of Lund, Sweden, in the 1950s, where a young orthopaedic surgeon was undertaking research into bone. Interested in gaining more insight into how bone fractures healed, Dr Per-Ingvar Brånemark (Figure 10.3) began by studying bone marrow cells in living animals.

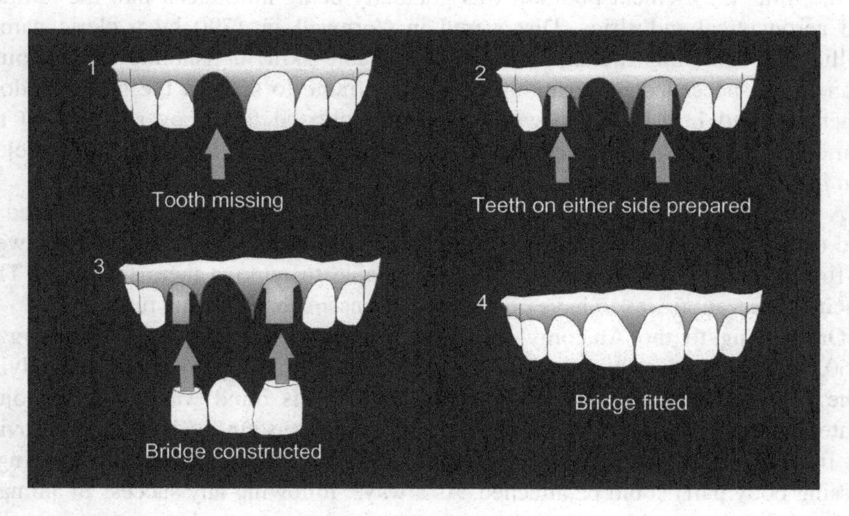

Figure 10.2 Diagram showing the principles of constructing a dental bridge. (See text for description.)
Source: Redrawn from P.-I. Brånemark et al., 1977.

Figure 10.3 Dr P.-I. Brånemark.
Source: Photographer Johan Wingborg.

He was familiar with techniques for observing blood vessels in the ears of live rabbits. These techniques involved using a microscope to look through a small glass window placed in the skin and supported by a frame constructed from a metal called tantalum.

Sensing that the same approach could be used for his own research, Brånemark set about developing a similar system. He removed the surface bone in the hind limb of rabbits to expose the deeper bone marrow and then covered the opening with a glass window supported by tantalum screws fixed into the bone. Through this window he was able to view the bone marrow with a microscope. At the end of the experiment, the expensive tantalum frame was easy to remove for reuse.

As he was not always certain of being able to obtain tantalum, Brånemark decided to use a different metal to act as a frame supporting the glass window. At the time, the element titanium was gradually being introduced into the military and aeronautical industries. Discovered in Cornwall in 1790 by a cleric named William Gregor, this metal required considerable skill in handling and shaping. Brånemark succeeded in designing a titanium frame to support the glass window. When inserted into bone, he was particularly pleased to discover how well the frame was tolerated, and noted the absence of any signs of inflammation or rejection in the rabbit tissue around it.

A few months later, at the end of the experiment, Brånemark was surprised to find that, unlike his previous experiments using tantalum, the titanium frames were so firmly bonded to the bone that he was unable to remove them for reuse. This crucial observation was to have considerable consequences for his future.

On moving to the Anatomy Department at the University of Gothenberg in 1960, Brånemark set up a multidisciplinary research group to continue studying bone from many different aspects. Uppermost in his mind was a major project related to titanium for if it bonded to bone and could sustain loads, it might provide the first successful material to which a prostheses (an artificial device replacing a missing body part) could be attached. As always, following any success in animals, it would then be necessary to replicate the studies in humans.

Critically, Brånemark demonstrated that bone formed directly against the titanium surface without the development of an intervening fibrous layer (Figure 1C).

He termed this unique feature 'osseointegration', which is mainly due to the presence of an incredibly thin oxide layer on the surface of the metal. When a titanium implant is inserted into bone, many important bioactive molecules (e.g. growth factors) accumulate at its surface and attract new bone-forming cells to bond directly with the titanium.

Among the many challenges faced and conquered by Brånemark and his team over this period were the ability to:

1. Engineer high-quality, pure titanium implants of the correct composition and shape
2. Surgically produce a channel in the bone with the closest fit between the implant surface and adjacent bone with minimum trauma
3. Operate in conditions of complete sterility

As an orthopaedic surgeon, Brånemark would have preferred the first experimental implants to be used in areas of his own expertise, such as the arm and leg. However, exposing the bone in these sites would have involved deep invasion of the tissues and potential damage to muscles, nerves and blood vessels. Fortunately for the dental profession, a small channel drilled in the jawbone for a dental implant presented none of these difficulties. Indeed, access is very easy, and once the gum is reflected, the bone is immediately exposed.

The first major scientific paper using titanium as a potential implant material appeared in the *Scandinavian Journal of Plastic and Reconstructional Surgery* in 1969. It was entitled 'Intra-osseous anchorage of dental prostheses' and described the successful retention of approximately 90 implants in dogs for up to 4 years. The basic procedure is a two-stage operation still used today and is shown in Figure 10.4. Starting from a situation where a tooth has been lost (Figure 10.4A), the first stage (Figure 10.4B) requires the bone to be exposed in the mouth, a channel to be drilled and a cylindrical titanium screw-thread implant inserted into the jawbone. The overlying gum is then replaced and sutured and the implant left to heal for about 3 months free from loading forces. After this period, wound healing results in the implant becoming firmly bonded (osseointegrated) to the adjacent bone (Figure 10.4C). The implant is then re-exposed and a post (abutment) inserted

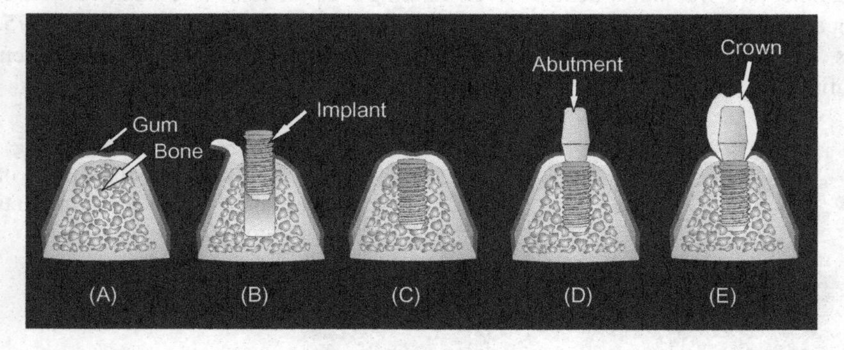

Figure 10.4 Line diagram showing principles of osseointegration. (See text for description.) *Source*: Redrawn from P.-I. Brånemark et al., 1977.

into a channel at the top of the implant (Figure 10.4D). Finally, a crown is fitted to the abutment (Figure 10.4E) to complete the surgical process. All of these procedures are of course carried out under local anaesthesia and under conditions of complete sterility. The adjacent tissues of the patient form a smooth attachment to the abutment without any signs of inflammation. Figure 10.5 shows an X-ray of three completed implants in a patient.

Instead of just a single implant, Brånemark's initial animal series involved four to six implants in each jaw to which was attached a single bridge containing a number of teeth.

The first patient to receive a Brånemark implant was Gösta Larsson. He had no teeth in the lower jaw, which was also deformed. In addition, he had a cleft palate. He was unable to chew properly, suffered considerable pain and had a poor quality of life. In 1965, with no other treatment on offer and nothing to lose, he was happy to volunteer to receive titanium implants. Four implants were inserted into his lower jaw to which, a few months later, a bridge with a titanium base supporting 12 teeth was fixed by screws. The operation was successful and the patient's lifestyle completely changed for the better. Larssen subsequently had his cleft palate treated and dental implants successfully placed in his upper jaw. He died in 2006 aged 75, still with successful implants in his jaws 40 years later.

Brånemark's first group of patients all had at least one jaw without any teeth (edentulous), many of them having difficulty in managing with traditional removable dentures, which adversely affected their quality of life. The basic technique was similar to that developed in animal studies. In each patient, four to six titanium implants were screwed into one or both jaws. After a period of healing, the implants were exposed and a bridge of titanium surmounted by 12 teeth was fixed to the implants of each jaw (Figure 10.6). Carefully recording all cases, in 1977 Brånemark was able to publish one of the most comprehensive reports of any clinical trial. This covered a 10-year period and was supported by results from over 2000 implants, with a success rate of over 90%.

Although the future of implants seemed all plain sailing, this was far from the truth. People were not ready to believe the evidence put before them, particularly as Brånemark was not a dentist. Even funding was difficult to obtain, but the evidence was so overwhelmingly positive that it could not be ignored, and in 1975, it was accepted by the Swedish Health Service as a legitimate form of dental treatment.

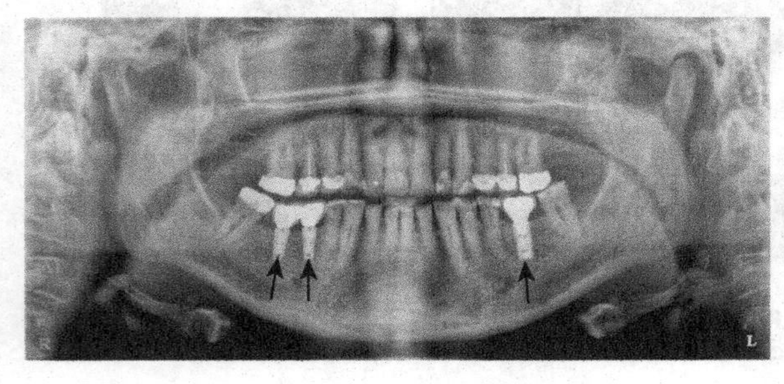

Figure 10.5 X-ray showing successful osseointegration of three titanium dental implants (arrows).
Source: Courtesy of Dr B.J. Doherty.

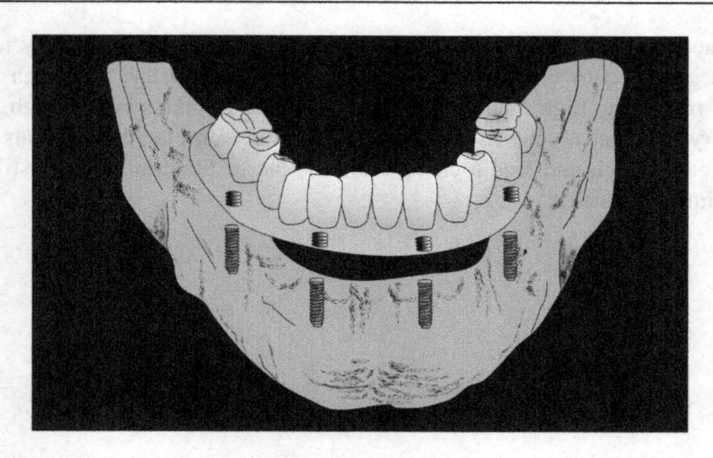

Figure 10.6 Diagram showing a fixed bridge in the lower jaw containing twelve teeth. The bridge is fixed to four titanium implants. Redrawn from Brånemark, P-I. et al., 1977.

On a worldwide basis, the clinical potential of a successful dental implant was enormous, with millions of suitable patients lining up for the procedure. However, a favourable outcome depended on adhering strictly to the clinical procedures defined by Brånemark and his team. To ensure his discovery was not tarnished by incorrect practise, Brånemark insisted that no surgeon could use his implant without attending his rigorous training course and without submitting their results to his archive. An engineering company was established to produce high-quality titanium implants, and a number of surgeons trained to fit them, first in Sweden and subsequently in Canada and the United Kingdom.

Fifteen years after the first patient received a titanium implant, none had been fitted in the United States. Perhaps there was some scepticism that a Swedish orthopaedic surgeon had found the answer to a fundamental problem which they had been unable to solve. Perhaps it was also because the results had not been published in a journal widely read by the American dental profession. Whatever the reasons, a symposium finally took place in Toronto in 1982 to which many American practitioners were invited. By the end of the conference, the case for the titanium implant was accepted and the stage set for its propagation worldwide. No one at this meeting could argue against the well-documented facts of over 15 years' clinical experience.

Since the first implant in a patient in 1965, millions of similar procedures have been carried out, from single-tooth replacements to whole dentitions, revolutionising dental treatment. There have been many new developments in implant design and construction. Nowadays, implants can be fully fitted in a single visit. In addition to fixed bridges, full dentures can be attached to implants by a ball-and-socket type joint so that they can be readily removed, the implants inspected and the denture replaced.

A chance observation made during a study on bone marrow cells by an orthopaedic surgeon led to one of the great advances in dentistry. Although the cost related to producing high-quality titanium for dental implants is still high, perhaps at this very moment a scientist working on a totally unrelated topic may make a similar breakthrough and find a cheaper alternative to titanium, though still retaining its unique properties.

11 Hens' Teeth Are Not as Rare as You Think

If something is said to be 'as rare as hens' teeth', it implies it is so rare it does not exist. This is because all the hundreds of species of modern-day birds, including hens, use a hard beak instead of teeth to collect food. One hundred and thirty million years ago, some of their extinct ancestors did have teeth (Figure 11.1), but around 80 million years ago, modern birds emerged without teeth.

In 1983, the late Professor Stephen Jay Gould, one of the great communicators of popular science, published a book of scientific essays entitled *Hen's Teeth and Horse's Toes*. The first part of the title, 'Hen's Teeth', was the subject of an essay in which Gould described experiments where the developing mouth tissues of chicks were manipulated to show that they still retain some potential to develop teeth. In the subsequent 30 years, more research has been carried out on this conundrum, which is the subject of this chapter.

Normal Tooth Development

For readers to better appreciate the topic, the normal development of teeth in mammals can be conveniently divided into five stages, as seen in Figure 11.2. At Stage 1, before the teeth have even started to develop, two distinct tissues can be seen lining the mouth. There is a thin surface layer of cells called the *oral epithelium* that covers a thicker underlying zone called the *dental mesenchyme*. The cells in each layer are all more or less similar in appearance.

The first evidence of tooth development occurs at Stage 2 when the surface oral epithelium invaginates into the underlying dental mesenchyme and, at certain sites, enlarges to map out the position of each future tooth. As development proceeds to Stage 3, there is a gradual change in each of the two layers, with an increase in the complexity of shape and structure. The epithelial invagination is now called the enamel organ and consists of three distinct layers. (A fourth layer appears later as it goes on to form enamel, which covers the outer surface of the tooth.) The deeper, dental mesenchyme that lies directly beneath the oral epithelium also increases in complexity of shape and is now called the dental papilla. It will later form the dentine (and dental pulp).

With further development at Stage 4 (see Figure 11.2), the cells of the enamel organ start to form enamel and the cells of the dental mesenchyme begin to form

Nothing but the Tooth. DOI: http://dx.doi.org/10.1016/B978-0-12-397190-6.00011-0

Figure 11.1 Reconstruction of an extinct species related to birds, Iberomesornis (meaning intermediate Spanish bird), that lived about 130 million years ago. Notice the teeth in its jaws.
Source: http://en.wikipedia.org/wiki/File:Iberomesornis-model.jpg. This work has been released into the public domain. This applies worldwide.

dentine. The major differences in structure between the future enamel and dentine (and the reason they stain differently) reflect the fact that the two tissues are derived from two different layers and have different compositions (see also Chapter 7). Once the crown has fully formed, at Stage 5 the root will develop and the tooth will erupt into the mouth.

The complex changes that occur in the original simple cell layers of the covering epithelium and underlying dental mesenchyme to eventually produce a tooth are due to the production and release of many complex signalling molecules (such as growth factors). These molecules pass back and forth from one layer to the other during development to effect changes by switching genes on and off (epithelial/mesenchymal interactions), eventually resulting in cells capable of producing enamel and dentine. The signalling process is fundamental to the successful development of all vertebrate animals. For a tooth to develop, both epithelial and mesenchymal layers must be present and both must be able to produce all the necessary complex biochemical molecules. This has been confirmed by experiments in which the two layers are separated early on in development and each grown in isolation when it is seen that further development into a tooth ceases.

Having briefly described the general principles concerning the interactions between the oral epithelium and underlying dental mesenchyme in forming a tooth, the question of why a hen doesn't have teeth can be addressed by looking at early development in its embryo, the chick.

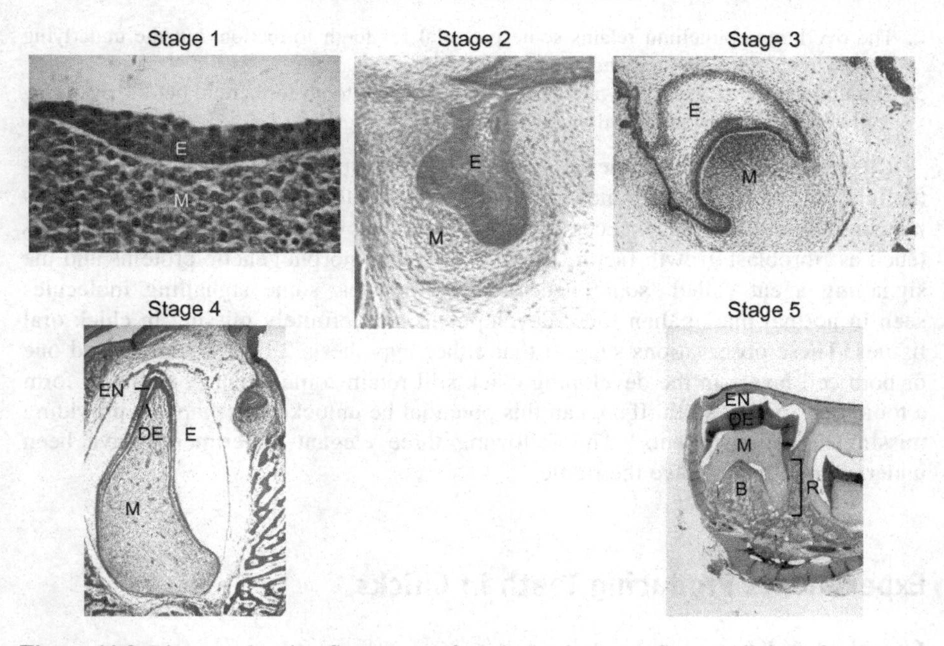

Figure 11.2 Diagram showing five stages of tooth development. Stage 1: lining of the mouth with two simple layers. E, epithelial layer; M, dental mesenchymal layer. Stage 2: invagination and initial swelling of the epithelial layer (E) into the underlying dental mesenchyme (M) to show the first signs of a tooth. Both layers are still composed of simple and similar cells. Stage 3: the epithelial layer now has a more complex shape and more discernable layers forming the enamel organ (E), which encloses the underlying dental mesenchyme cells (M), now called the dental papilla. Stage 4: the epithelial enamel organ (E) showing more complexity of shape and commencing to form the enamel layer (EN) at the tip of the cusp. It surrounds the mesenchymal dental papilla (M), which has started to form the dentine layer (DE) beneath the enamel. Stage 5: the crown has been completed and the root starts to form, eventually resulting in the tooth erupting into the mouth. EN, enamel space; DE, dentine; M, dental pulp; B, bone of socket; R, root.
Source: From B.K.B. Berkovitz, G.R. Holland and B.J. Moxham, 2009. *Oral Anatomy, Histology and Embryology*. 4th edition. Elsevier.

Why Teeth Do Not Normally Develop in Chicks

The mouth of the chick has the same two developing layers as any mammal, namely the surface epithelium and the underlying mesenchyme. As chicks never develop teeth (although its distant ancestors had teeth), three main hypotheses can be proposed to explain this absence:

1. Both the epithelium and mesenchyme have lost all potential to form teeth, with both layers lacking specific genes for producing all the appropriate signalling molecules: double incompetence.

2. The overlying epithelium retains some potential for tooth formation, but the underlying mesenchyme has none: mesenchymal incompetence.
3. The underlying mesenchyme retains some potential for tooth formation, but the overlying epithelium has none: epithelial incompetence.

Regarding the first hypothesis, studies have shown this is not true. Both the epithelium and the underlying mesenchyme in the mouth contain some of the important signalling molecules necessary for normal tooth development in mammals (such as fibroblast growth factor, some of the bone morphogenetic proteins and the signalling agent called 'sonic hedgehog'). However, some signalling molecules seen in normal mammalian tooth development are definitely missing in chick oral tissues. These observations suggest that either hypothesis 2 or 3 is correct and one or both cell layers in the developing chick still retain some latent potential to form a tooth or part of a tooth. If so, can this potential be unlocked, perhaps by providing missing signalling agents? The following three elegant experiments have been undertaken to try to solve the riddle.

Experiments Producing Teeth in Chicks

Experiment 1

Because some important signalling molecules are absent in the mouth epithelium and mesenchyme of the chick, this experiment attempts to reintroduce them during development.

Although mesenchyme beneath the mouth epithelium in the chick lacks the important growth factor, bone morphogenetic protein 4 (BMP4), it has been shown that the mesenchyme in the developing skin of the chick does produce this missing growth factor. It is possible to separate the two layers in the mouth and skin and recombine the oral epithelium in the mouth with the mesenchyme from the skin of the chick (Figure 11.3). The two tissues can then be grown in a test tube. This experiment is the equivalent of exposing the mouth epithelium to the missing growth factor BMP4. Amazingly, when this experiment was carried out, the mouth epithelium combined with the skin mesenchyme to produce a recognisable tooth bud (although it was unable to develop any further and produce mineralised tissue). This is the first example of the existence of a hen's tooth (Figure 11.4).

Experiment 2

It has been found possible to mix chick tissue with that of the mouse without tissue rejection. In this second experiment, the normal chick mesenchyme is replaced by mesenchyme from a developing mouse tooth, which obviously has all the signalling factors necessary to form a tooth. In this way, it is possible to determine whether the normally unresponsive chick epithelium retains any capacity to form a tooth with more appropriate stimulation from mouse dental mesenchyme.

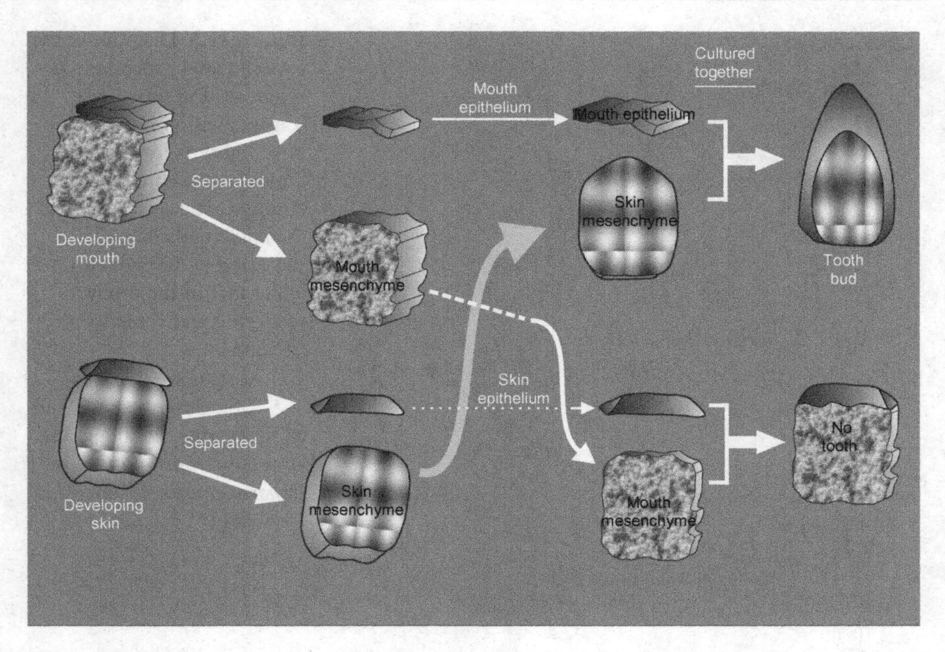

Figure 11.3 Diagram showing separation of epithelium and mesenchyme from the mouth and skin and recombining them in different combinations. When the mouth epithelium is cultured with skin mesenchyme a developing tooth is produced.

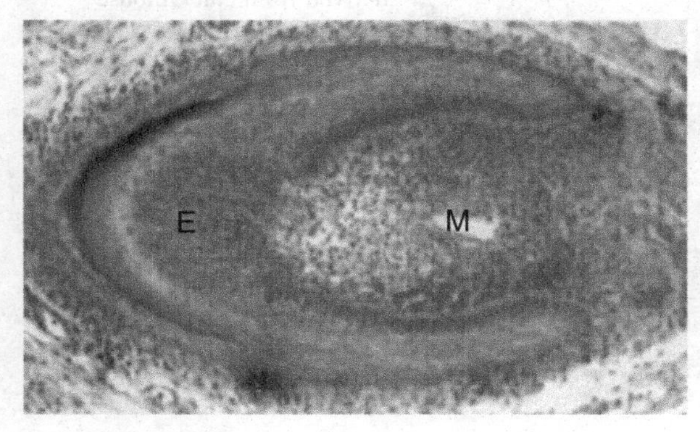

Figure 11.4 Chick 'tooth-like structure' formed by recombining chick mouth epithelium (E) with chick skin mesenchyme (M) and culturing for 2 days.
Source: From Y. Chen et al., 2000 and the editors of *Proceedings of the National Academy of Sciences of United States of America.*

During very early development, the cells that form the mesenchyme in a mouse tooth are initially located in the developing brain region, from which they migrate into the jaws. For the second experiment, these cells are dissected out from the developing mouse embryo and exchanged for similar cells in a chick embryo developing in an egg (Figure 11.5). The resultant chick continues to develop with a mixture of cells derived from both itself and a mouse. Using special dyes, it is possible to readily distinguish between the two sources of cells. The resulting mouse/chick

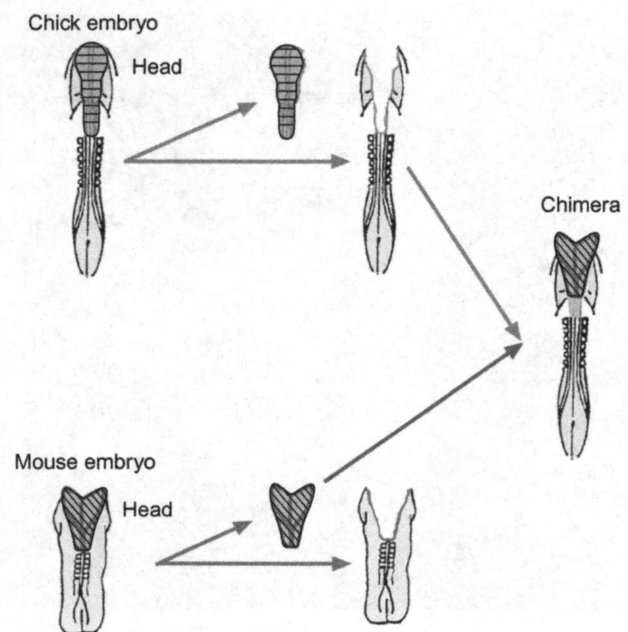

Figure 11.5 Diagram of the experimental procedure of mouse neural tube transplantation into chick host embryos for the creation of a chick/mouse chimera. The neural tube of the chick embryo is indicated by horizontal lines while that of the mouse embryo is indicated by oblique lines.
Source: Courtesy of T.A. Mitsiadis, J. Caton and M. Cobourne, 2006 and the editors of *Journal of Experimental Zoology* (*Mol. Dev. Evol.*).

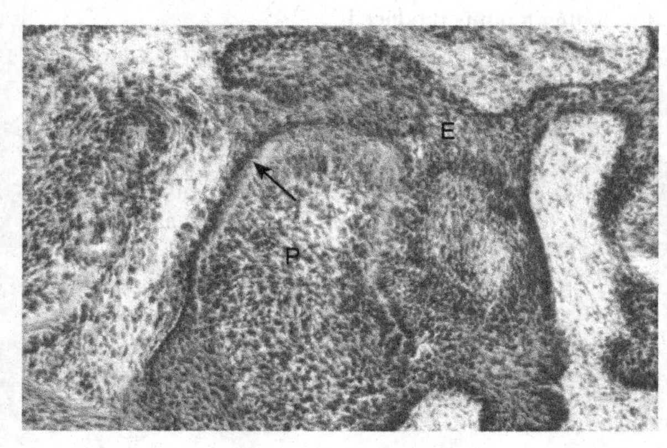

Figure 11.6 Chick tooth derived from chick/mouse chimera experiments. The epithelial enamel organ (E) is chick derived and the dental papilla (P) is mouse derived. The staining (arrow) represents initial dentine matrix.
Source: Courtesy of Dr T.A. Mitsiadis.

chimeras will produce tooth germs. The special dyes indicate that the outer, epithelium component of the tooth is clearly produced by the chick epithelium and the inner dental mesenchyme is clearly produced by the mouse cells. This type of tooth also reached a more advanced stage of development than that seen in experiment 1, even producing some proteins characteristic of dentine (Figure 11.6). This experiment is again a confirmation that chick epithelium still retains the potential to interact with suitable mesenchyme (in this case, mouse mesenchymal cells) and participate in tooth development. That these experimental teeth fail to mineralise completely to produce enamel is not unexpected as chick mouth epithelium has been shown to completely lack the genes necessary to produce enamel. Here then is our second example of a hen's tooth.

Teeth Occurring Naturally in Chicks

The previous two examples of hen's teeth have been produced experimentally by manipulating the mesenchymal components. In this third example of hen's teeth, teeth occur naturally, but in a chick mutant. This condition also sheds light on why the teeth appear and helps us to interpret the observations from the preceding two experiments.

Mutations occur naturally in all animals and are the result of abnormalities in the DNA sequence of the genes. As the DNA sequence determines the type of protein produced, mutations can result in a protein: (i) not being produced, (ii) being produced in too little a quantity, (iii) being produced in too great a quantity, (iv) being produced at the wrong time or (v) being produced with a different chemical structure. A mutation may have a wide range of effects, from very little or no effect to serious defects that may be lethal. The study of mutations has resulted in major advances of our understanding of both normal and abnormal development.

Large numbers of mutations have been detected in the chick, some occurring naturally and others having been produced experimentally. One such mutation occurring in the chick has been termed 'talpid2' (talpa = mole, as the abnormal limbs of the mutant chick have a spade-like appearance similar to that of moles). Because of disruptive changes in some important signalling molecules (especially in 'sonic hedgehog'), many abnormalities occur in the limbs and head and the chick dies before hatching. In the head, for example, the beak region is foreshortened compared with the normal chick (Figure 11.7). Although this mutant was discovered 50 years ago, until recently nobody had looked in detail at the developing beak region. Remarkably, when scientists did, they discovered a number of small, unmineralised conical tooth germs (Figure 11.7). In this mutant chick, therefore, developing tooth germs are present as a normal feature, although they do not progress to a later stage of mineralisation.

Explanation of Why Normal Chicks Lack Teeth

How is it possible to explain the appearance of teeth in the talpid2 mutant chick but not in the normal chick? The scientists involved in the discovery carried out a

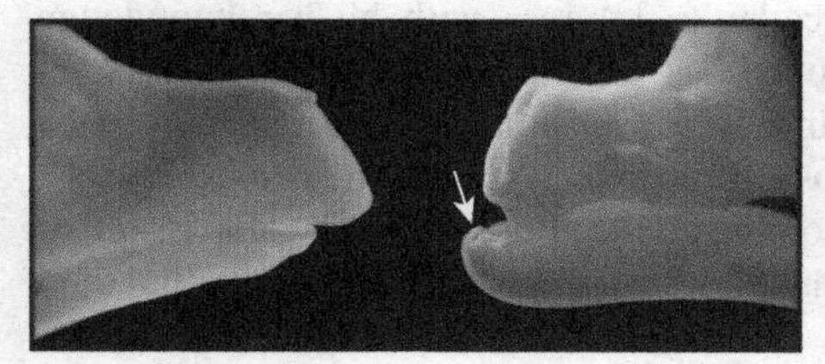

Figure 11.7 Normal chick embryo on left. Talpid2 mutant on right, showing small teeth projecting from the beak region (arrow). Note the abnormal growth with foreshortening of the beak in the mutant. *Source*: Courtesy of Dr M.P. Harris.

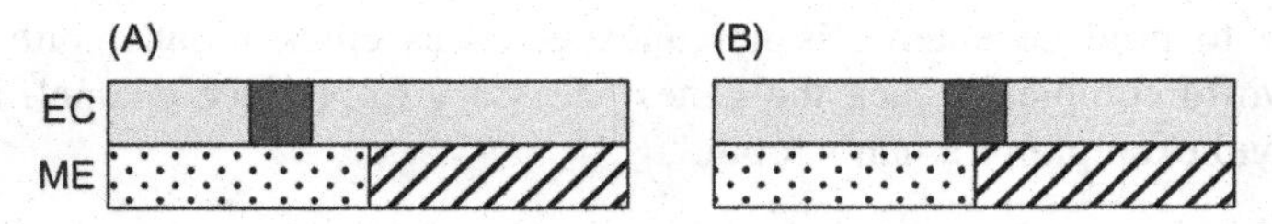

Figure 11.8 Diagram showing the alteration in the position of the tooth-producing mesenchyme in normal and talpid2 mutant (ta2) jaw, leading to the initiation of teeth in the mutant. (A) In the normal chick, the responsive area of the oral epithelium (EC) is indicated in solid black. The responsive underlying dental mesenchyme (ME) is indicated in diagonal lines and is not directly beneath the oral epithelium but is shifted to one side. (B) In the mutant chick, abnormal growth patterns leads changes in the relative positions of the two layers such that the epithelial signaling centre (solid black) and underlying competent mesenchyme (diagonal lines) are juxtaposed, permitting initiation of tooth development. *Source*: Courtesy of M.P. Harris et al., 2006 and the editors of *Current Biology*.

detailed study of the signalling molecules present in the developing mouth region in both normal and mutant specimens. They confirmed that in the normal chick, the oral epithelium does contain tooth-promoting factors. They also noted that the mesenchyme immediately beneath the epithelium, while containing some signalling molecules, lacks others needed for tooth development to commence. However, they additionally found that adjacent mesenchyme cells to the side also contain signalling molecules that have the potential to produce small tooth germs, but these cells are too far away to interact with the epithelium. Hence, normal chicks cannot develop teeth (Figure 11.8A). However, in the mutant chick, the researchers discovered that the disturbed growth processes producing the severe cranial and facial deformities (Figure 11.7B) had fortuitously resulted in displacement of the tooth-potentiating mesenchyme so that it came to lie directly beneath the oral epithelium (Figure 11.8B). Instead of being offset, the two components necessary for tooth formation could now interact and produce the third example of hens' teeth in the talpid2 mutant (Figure 11.7B).

Summary

This account has shown three different instances of hens' teeth, two experimentally produced and one occurring normally (but in a mutant). The statement 'as rare as hens' teeth' has lost some of its meaning, at least to the few who know about these things. Perhaps it will also enable the reader to impress with his/her knowledge should the opportunity arise when someone uses the phrase. In addition, research into the topic has provided an important insight into evolutionary mechanisms, as a major evolutionary change (such as the loss of a tooth or teeth in a species) doesn't require an impossibly difficult explanation. It could simply arise as the result of a minor change in the relative positions of normally interacting epithelial and mesenchymal cell layers.

12 John Hunter and the London Tooth Museum

The heart of Leicester Square in the West End of London is pedestrianised and contains a small park, Leicester Square Park. In its centre are statues of two of England's most famous entertainers, one ancient and one more modern. The larger statue is of the playwright William Shakespeare, while next to it is a statue of the silent screen comedian, Charlie Chaplin. At each corner entrance to the park is a bust representing a famous Englishman from the eighteenth century. Three are well known, namely Sir Isaac Newton, mathematician and astronomer (1643–1727); Sir Joshua Reynolds, portrait painter and one of the founders and first president of the Royal Academy of Arts (1723–1792) and William Hogarth, painter, engraver and satirist (1697–1764). The statue on the remaining corner is of John Hunter (Figure 12.1), surgeon and scientist, who is less well known and the subject of this chapter.

John Hunter and His Contribution to Medicine and Science

John Hunter was born in 1728 in Scotland, the 10th and youngest child of the family. Brought up on a small farm in East Kilbride and showing little promise at school, his career prospects did not look good. Realising the need for change if he were to achieve anything, when he was 20 years old he wrote to his elder brother, William, offering his services in any capacity.

William Hunter, 10 years older than John, had studied medicine in Edinburgh before moving to London in 1741 as a young trainee surgeon. In those days, physicians enjoyed high status and rarely deigned to examine patients physically. Surgeons, on the other hand, did not have such a high standing. Before the days of anaesthetics and antisepsis, and with little detailed knowledge of anatomy, only minor surgical procedures had any chance of success. Surgical work such as blood-letting or tooth pulling was often left to the barber-surgeons who themselves had little, if any, training.

Determined to transfer to the more lucrative field of obstetrics, the ambitious William Hunter realised the importance of knowledge of human anatomy both in medical practise and in helping to improve success rates in surgery. With this in mind, in 1746 he established a private school of anatomy in Covent Garden. The key to success for his venture was to guarantee every student would have the

Nothing but the Tooth. DOI: http://dx.doi.org/10.1016/B978-0-12-397190-6.00012-2

Figure 12.1 Author by statue of John Hunter in Leicester Square Park. The park is undergoing reconstruction (as of April 2012).
Source: Photographed by Mr G. Fox.

opportunity to dissect human bodies. This meant that he had to overcome the problem of legally obtaining sufficient numbers of bodies.

John's arrival in London in 1748 fortuitously coincided with the beginning of anatomy classes for a new intake of medical students at his brother's fledgling anatomy school. Help was needed in providing students with guidance in human dissection, and William offered his young brother employment as a dissector. The uncomfortable conditions under which John found himself working, particularly regarding the smell of the corpses and the very cold surroundings, can only be imagined, not to mention his total ignorance of the subject. As the science of preserving cadavers was still in its infancy, dissection had to be carried out in the cold winter months in order to delay the rate of decomposition of the bodies.

In a relatively short time, it became clear that John had a remarkable aptitude for dissection. He also possessed a scientific curiosity and talent for developing methods of preserving and displaying teaching material, especially of diseased structures. John demonstrated initiative in procuring sufficient corpses, even if this meant undertaking grave-robbing himself and dealing with the 'resurrectionists', who sold bodies without too many questions being asked about their origins.

Within a short space of time, it was apparent there was little that William could teach John, who was soon in charge of the dissecting room. From that moment on, John Hunter found his purpose in life. His hidden and considerable intellect began to flourish. From humble beginnings and lack of education, he eventually rose to become the most famous anatomist and surgeon in England.

John Hunter's reputation as an anatomist quickly grew. His skill in dissection was greatly admired, and he started his own research into the workings of the human body, little of which was known at the time. Soon, medical students from around the world were coming to London to learn anatomy under his tutelage. When these students returned home, they took with them Hunter's ideas and, most importantly, his meticulous scientific approach.

Among John Hunter's early major scientific discoveries, two can be highlighted, either of which would have assured him a place in the history of anatomy. First was his demonstration that the lymphatic system was separate from the blood system. Second was his discovery that the blood circulations of a human foetus and its mother were separate and not joined directly together, as was the general opinion at the time.

John undertook dissections of human foetuses at the behest of his brother William, who was planning a book containing a small number of illustrations showing the development of the human foetus. This he hoped would establish his reputation in the field of obstetrics. When William saw his brother's beautiful dissections, he decided to postpone publication until he had a larger, complete series of specimens. This project was not finished for 25 years when his atlas *The Anatomy of the Human Gravid Uterus Exhibited in Figures* was finally published in 1774 (Figure 12.2). This book is considered one of the most important in the history of medicine. However, John Hunter felt that he was not given sufficient credit, particularly for establishing the separateness of the foetal circulation. A bitter disagreement erupted into the public domain, and, sadly, the two brothers barely spoke

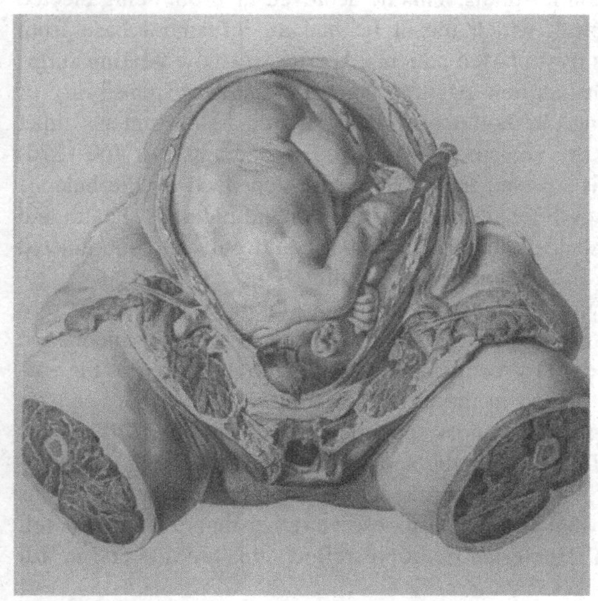

Figure 12.2 Illustration of a human foetus dissected by John Hunter and illustrated by Jan van Rymsdyk in William Hunter's book: *The Anatomy of the Human Gravid Uterus Exhibited in Figures*. *Source*: Reproduced by kind permission of the Hunterian Museum at the Royal College of Surgeons.

to each other again. The artist, Jan van Rymsdyk, who produced all the wonderful engravings, was given no recognition at all.

Having acquired more anatomical knowledge than anyone else in the country, John Hunter was aware that this alone would not help his advancement. He chose to pursue a career in surgery and left his brother's anatomy school. He spent the next 3 years as an army surgeon, which required little previous experience to enlist. This took him to France and Portugal, where he gained experience in treating many war casualties. He returned to London in 1763, where his army service enabled him to establish a practice as a surgeon. Although the next few years were financially difficult, Hunter's surgical and scientific reputation spread and by 1768 he was sufficiently experienced to be accepted into the elite Company of Surgeons.

At this time, autopsies were being undertaken (mainly at the behest of the legal profession) to help determine not only the cause of death but also to aid the medical profession in improving diagnosis and treatment. Hunter was soon acknowledged as the most experienced person in England to carry out these postmortems.

Hunter continued his experiments on a wide range of topics on animals while continuing to dissect and preserve human material, both normal and abnormal. His aim was to try and understand, through his own observations and experiments, how the body functioned. He was not prepared blindly to accept existing medical theories, techniques and treatments that had been passed down dogmatically from bygone ages, particularly where his own experience indicated these were incorrect.

Hunter's career as a surgeon could only advance significantly if he obtained a post in one of the major London hospitals. This he achieved in 1768, being elected surgeon at St George's Hospital, which meant he had an influential base from which to work. Apart from his own private patients, Hunter was now treating surgical cases in his hospital. With this new position came the added responsibility of supervising surgeons-in-training. It should be remembered that in Hunter's time, medicine was still following the ancient doctrines of Hippocrates (BC 460–370) and Galen (129–200 AD). This stressed that good health depended on the balance of the four basic humours: black bile, yellow bile, phlegm and blood. Ill health and disease resulted from an imbalance between them that could only be counteracted by procedures such as purging the patient or bloodletting.

Hunter introduced the study of comparative anatomy, looking at a wide range of animals to observe how structure and function were related in order to understand the workings of the human body. His lectures and demonstrations were not the easiest to follow, and he was not gifted in the art of speaking. However, he used his unrivalled and ever-expanding collection of specimens to educate and inspire his growing numbers of students from both home and abroad. It was through these students that his philosophy of careful experimentation, observation and rigorous thought was spread: 'Believe what you actually observe rather than what others tell you should happen'. The term 'Hunterian tradition' reflects this attitude and is still used today.

One of Hunter's students was Edward Jenner. With encouragement from Hunter, Jenner founded the science of immunology through the successful inoculation of patients with cowpox to prevent the onset of the related, but far more dangerous disease, smallpox.

Hunter's scientific eminence was recognised when he was elected a Fellow of the Royal Society in 1767. He built a large house at Earls Court with facilities to dissect and accommodate his ever-expanding collection of specimens, later moving to much bigger premises in Leicester Square. This was still a well-known building in the 1880s, and author Robert Louis Stevenson is thought to have used it as a model for Dr Jekyll's house, with its laboratory, in his novel *The Strange Case of Dr Jekyll and Mr Hyde*'.

Hunter was continually on the lookout for unusual specimens showing abnormalities, hoping they would help him gain a deeper understanding of how organs and individuals developed. He was known to dealers as willing to pay high prices for such material. Due to the ongoing expense of obtaining, preserving and housing his ever-expanding collection, Hunter was always in need of funds, despite enjoying a considerable income as one of the major London surgeons, plus fees from students. His pre-eminence in surgery was recognised by his appointment in 1776 as Surgeon Extraordinary to King George III.

No object was too large or too small to escape Hunter's interest. He obtained the skin and some bones of the first giraffe ever to be seen in England and dissected a bottle-nosed whale, whose skeleton he displayed. He received and described some of the first marsupials following the discovery of Australia by Captain Cook. He received the corpses of many exotic animals that died in the private zoos of the wealthy, including those from the royal menagerie at the Tower of London.

One story relating to Hunter's desire to possess unique specimens concerned Charles Byrne, the Irish Giant, who, in 1782, caused a sensation when he appeared in London at the age of 21. At that time, there was a great demand for 'freak shows', which included the display of giants and dwarves. Byrne was about 2.3 m tall (7 ft 8 in), the result as we know today of a tumour in his pituitary gland that caused the production of excessive amounts of growth hormone. Unfortunately, it soon became apparent that, due to the severe side effects of the untreated tumour, Byrne's health was rapidly deteriorating and that he would soon die. It was also no secret that Hunter, among many other competing individuals, wanted this body to dissect and add to his collection, and would be willing to pay a large sum for the privilege. Terrified of being dissected, Byrne invested all the money he had, which was no small amount, into what he regarded as a foolproof scheme that would prevent his body getting into the hands of the anatomists. His plan was to hand over his money to an undertaker with strict instructions that he be interred in a large, heavy coffin watched over by loyal friends. These friends would then transport him to the coast, transfer him to a boat and bury him deep at sea. Following Byrne's death in 1783 at the age of just 22, this request was seen to be carried out. However, somewhere along the route Byrne's body 'disappeared' (assuming it was in the coffin in the first place!) so that the coffin buried at sea did not contain his remains. Five years later, in 1788, when Hunter first opened his collection of specimens to a select group of friends and colleagues, Byrne's skeleton took pride of place and was there for all to see, its enormous height causing a sensation. It was not clear whom Hunter bribed, or how the body was obtained, but a very large sum of money must have changed hands.

Throughout his life Hunter continued to produce a stream of major scientific papers. Among important observations were those showing how bones increased in size. To shed light on this mechanism, he designed a simple experiment whereby he inserted two small metal pellets along the shaft of the leg bone of a young chicken, initially recording their distance apart. After a period of time, when the bone had grown, he found that the two metal pellets were still the same distance apart. This simple but elegant experiment proved that the bone had increased in size by growth at the ends of the bones rather than in the central shaft (Figure 12.3). This result he further confirmed by adding a natural dye called madder to the diet of animals, which stained newly formed bone red. Never afraid to experiment upon himself, Hunter also swallowed some madder and noted that it discoloured his urine.

Hunter published two important books that furthered his scientific reputation even more, namely *A Treatise on the Venereal Disease* (1786) and *Blood, Inflammation and Gunshot Wounds* (1793). The first book was a very detailed account of the clinical features of venereal disease throughout its many stages.

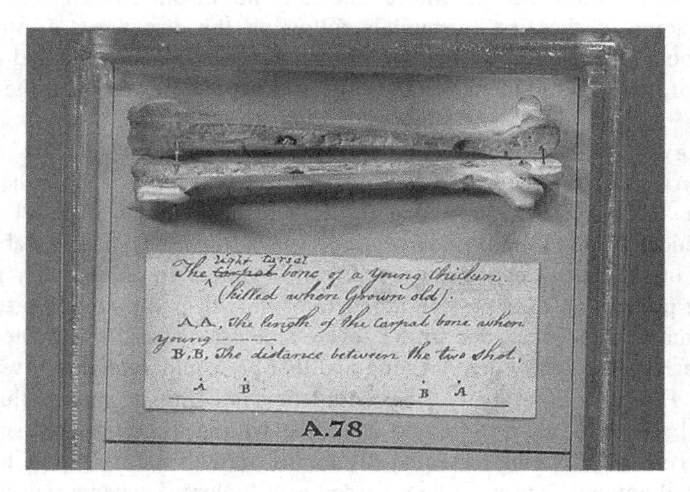

Figure 12.3 The main bone (tarso-metatarsus) in the leg of a chicken. John Hunter made two cauterised holes near the ends of the bone and filled them with metal. The chicken was later killed, and the holes were found to be of a similar distance apart, while the overall length of the bone had increased. This elegant experiment demonstrated that bone growth occurred in the ends (epiphyses), rather than in the shaft (diaphysis) of the bone. The original notes of the experiment made by Hunter's assistant William Bell accompany the specimen.
Source: Reproduced by kind permission of the Hunterian Museum at the Royal College of Surgeons.

Hunter describes inoculating some of his own patients with diseased tissue. There is even a belief that the patient from whom many of the accurate signs and symptoms were recorded was Hunter himself, and that he had deliberately inoculated himself to obtain the information. Although this cannot be confirmed, it could account for some of the health problems that pursued him in later years and that may have contributed to his death in 1793. His second book was related to his previous experiences as an army surgeon and considered many clinical aspects of gunshot wounds.

On 16 October 1793, during an acrimonious meeting with fellow surgeons at St George's Hospital, Hunter suddenly dropped dead. As befitted the man, he had requested that, on his death, an autopsy be performed to seek any underlying cause. This was carried out by his brother-in-law, Everard Hume, and revealed an unhealthy heart, which was consistent with the angina from which Hunter had suffered for many years. He was buried quietly in St Martin's in the Fields, London. In 1859, when this site underwent redevelopment, his body was removed and a belated, but fuller, acknowledgement of his enormous achievements was made by the reinterment of his body, with all due ceremony, in Westminster Abbey. On his tomb is inscribed the epitaph 'John Hunter – Founder of Scientific Surgery'.

On Hunter's death, his family was faced with poverty, for despite his considerable earnings, his major asset on which he spent most of his money was his incomparable collection of anatomical specimens. In 1799, the British government purchased the collection on behalf of the nation, albeit at a knockdown price, and the proceeds were given to his family. In 1800, the importance of surgery, for which Hunter had fought throughout his illustrious career, was recognised with a Royal Charter that founded the Royal College of Surgeons of London (later of England). With the construction of a new building in Lincoln's Inn Fields in 1806, Hunter's collection of anatomical specimens became the Hunterian Museum of the Royal College of Surgeons of England.

The Royal College of Surgeons is responsible for maintaining, overseeing and improving standards of surgery in England. This is principally achieved through its educational and examining programs. Generations of surgeons have passed through its corridors, visiting and admiring the contents of the Hunterian Museum. The collection has expanded significantly over the years from Hunter's original 13,500 specimens to well over 65,000. Conservators of the stature of Sir Richard Owen (who coined the term 'dinosaurs') and Sir Arthur Keith (of Piltdown fame; see page 167–170) have continued to care for the museum. Even today, Hunter's presence in the college is still strong. In addition to the magnificent museum, an annual Hunterian lecture continues to be given, and it is traditional for newly qualified surgeons to pose for a photograph by his statue, which dominates the entrance hall of the Royal College of Surgeons (Figure 12.4).

One sad episode of note is that during the Second World War, the Royal College of Surgeons of England was bombed and about two-thirds of the museum specimens were irretrievably lost.

Figure 12.4 Statue of John Hunter in the entrance to the Royal College of Surgeons of England.
Source: Reproduced by kind permission of the Hunterian Museum at the Royal College of Surgeons.

John Hunter's Contribution to Dentistry

The reader may be wondering why the founder of scientific surgery in England should be given such prominence in a book about teeth, with no reference so far to teeth having been mentioned. In collecting his animal material, Hunter could not ignore the skull and dentition, particularly as it most readily demonstrated the relationship between form and function, a theme close to his heart. The most famous portrait of Hunter, by his friend Sir Joshua Reynolds, depicts him contemplating a manuscript, with a book open on the table in front of him showing a series of skulls (Figure 12.5). Also, at the base of his statue in the entrance hall of the Royal College of Surgeons of England (Figure 12.4) a book is similarly open at a page containing animal skulls, with teeth prominently displayed.

Apart from his general interest in animals' teeth, Hunter developed a much more specialised knowledge of human teeth that established him as the founder of scientific dentistry as well as scientific surgery. This came about on his return from army service in 1763, when he found employment scarce.

In the eighteenth century, although primitive attempts at filling teeth were undertaken, the main treatment for dental problems was tooth extraction. Dental treatment had a low priority in the eyes of the medical profession and was mainly

Figure 12.5 Portrait of John Hunter painted by Sir Joshua Reynolds. Note the book on the table showing skulls and teeth. *Source*: Reproduced by kind permission of the Hunterian Museum at the Royal College of Surgeons.

the province of barber-surgeons, who had little training. Nevertheless, particularly skilled dental practitioners with greater knowledge and expertise were in demand by wealthy patrons, especially bearing in mind the lack of anaesthetics at that time (see Chapter 3). One well-known and respected London practitioner was James Spence. James Boswell, the renowned biographer of Dr Samuel Johnson, noted in his diary 'Toothache easier. Went to Spence. Two stumps drawn and teeth cleaned: agreeable to see thing well done'.

To provide Hunter with a source of income on leaving the army, he teamed up with James Spence, acting as a consultant and a support for Spence's practice. Hunter was all too aware of the lack in basic knowledge of the subject and soon set about the task of correcting this. He spent considerable time and effort in mastering the subject, adding to his collection of specimens numerous examples of teeth, both healthy and diseased.

Hunter undertook experiments on teeth, one of which was very far sighted for the period. It was not uncommon for a rich client who needed a tooth extracted to have it replaced by one removed from a servant. The transplanted tooth would have been held in place by ligatures, although its long-term prognosis was very poor. This led Hunter to develop an interest in tooth transplantation. In an initial series of experiments, he dissected out the spur from the leg of a chicken and transplanted it into the vascular tissue of the coxcomb in the head. In this site, the spur continued to grow, reaching a considerable size (Figure 12.6). Ever more

Figure 12.6 Head of a chicken showing the continued growth of a spur transplanted from the foot to the coxcomb.
Source: Reproduced by kind permission of the Hunterian Museum at the Royal College of Surgeons.

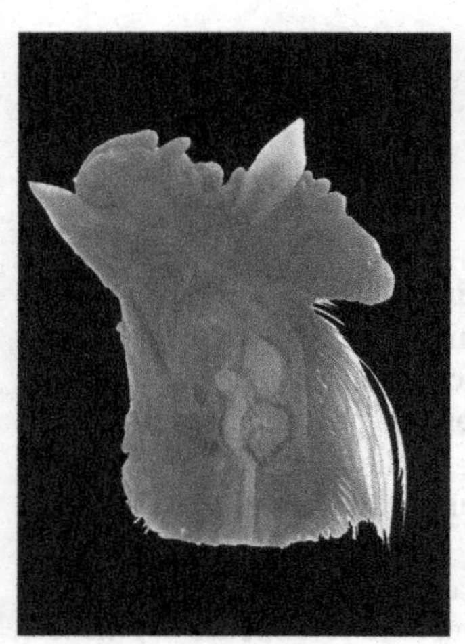

Figure 12.7 Head of chicken in which a tooth has been transplanted to the coxcomb.
Source: Reproduced by kind permission of the Hunterian Museum at the Royal College of Surgeons.

ambitious, Hunter transplanted human teeth to the same site in the hope that they would remain vital, perhaps providing a ready source of teeth for transplantation. A specimen from one such experiment is shown in Figure 12.7. Although seemingly a successful transplant, close inspection suggests that it had not been successfully grafted and would eventually have been rejected.

Dental studies by Hunter provided him with sufficient information to publish his first two books that not only helped his reputation as a scientist but also to establish dentistry as a serious scientific discipline. These two books, *The Natural History of the Human Teeth* (1771) and *A Practical Treatise on the Diseases of the Teeth* (1778), encompassed all that was known about the structure and development of teeth, diseases of the teeth and aspects of clinical treatment. Unlike his brother

William, Hunter acknowledged the artist Jan van Rymsdyk for the engravings. Together with Pierre Fauchard's earlier work in French in 1728, '*Le Chirurgien Dentiste*', Hunter's books were a major influence on the subsequent development of dentistry. Dental specimens illustrated in these books can be readily identified in the Hunterian Museum today.

Hunter's original collection of dental specimens was significantly increased when a large number of additional dental specimens (known as the Odontological Series) were incorporated into it from the museum of a separate dental organisation, the Odontological Society. Because of this and other specimens added since, the Hunterian Museum contains one of the most important dental collections in the world, with over 11,000 specimens. It is widely used by researchers from all over the world, and specimens are fully catalogued and available for viewing over the internet.

The Tooth Collection at the Hunterian Museum

In addition to the Irish Giant, the museum has the skeleton of Caroline Crachami, also called the 'Sicilian Fairy', one of the smallest people who ever lived. There is no information concerning her birth and infancy, but she was reputed to be 9 years old and only 53 cm (21 in) tall (Figure 12.8). She caused a sensation when first exhibited in London in 1824 and was even presented at Court to King George IV. Unfortunately, she died of tuberculosis only a few weeks after arriving but was well known enough in the capital to have an obituary in *The Times* newspaper. When she died, her 'guardian', a bogus doctor called Gilligan, absconded, leaving behind significant debts and was never heard of again. Caroline Crachami's skeleton is displayed together with some of her possessions, including a tiny ring and a pair of her shoes. Her particularly rare medical condition is known as microcephalic osteodysplastic primordial dwarfism, type 3, also known by her name as Caroline Crachami type.

Recent detailed examination of her teeth has led to the conclusion that she was not 9 years old when she died, but nearer 3. Presumably her age was inflated for commercial reasons as people might have been reluctant to pay to see a 3-year-old dwarf compared with one aged 9. However, even if age 3, this should not detract in any way from her incredibly small size. Some of her teeth show signs of tooth decay with bony abscesses, which must have caused her pain. Caroline Crachami has been the subject of many scientific articles and even a play, *The Smallest Person*, presented at the Edinburgh Festival in 2004.

The museum contains a series of skulls of children of various ages dissected to show how the teeth develop within the jaws and then erupt into the mouth (see Figure 4.12).

A large number of animal skulls can be viewed, showing a wide range of dentitions, from carnivorous fish with numerous sharp, pointed teeth that are continually replaced (see Figure 4.10) to old elephants that have only one remaining pair of

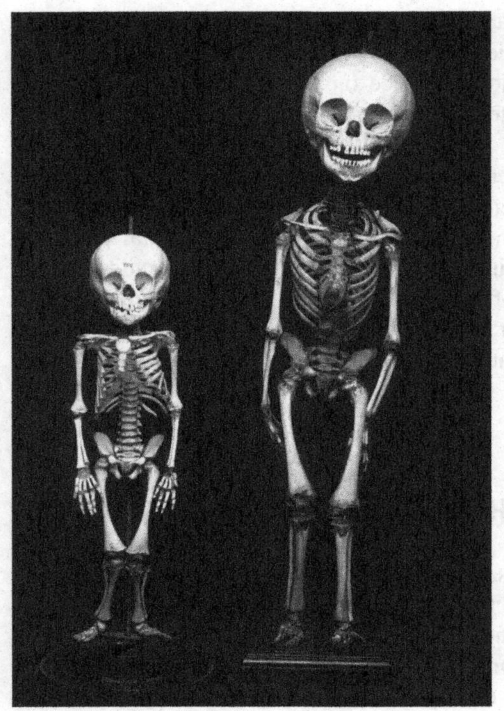

Figure 12.8 Skeleton of the dwarf Caroline Crachami on left compared with the skeleton of a normal 9-year-old child. *Source*: Reproduced by kind permission of the Hunterian Museum at the Royal College of Surgeons.

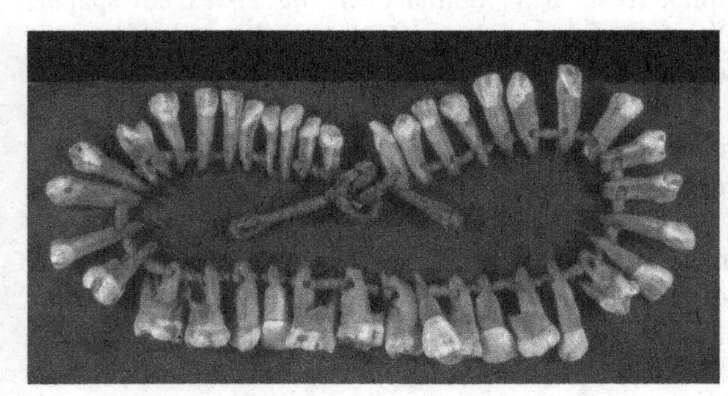

Figure 12.9 Necklace of human teeth presented to the Royal College of Surgeons of England by the explorer Henry Morton Stanley. *Source*: Reproduced by kind permission of the Hunterian Museum at the Royal College of Surgeons.

cheek teeth on either side of the jaw to grind on, after which the animal will die. The collection contains fossil teeth, including those of the largest shark ever known, Megalodon (see Figure 4.5).

The great explorer Henry Morton Stanley led a number of expeditions into Africa between 1871 and 1889, during the first of which he searched and tracked down another famous explorer, Dr Livingstone, who had completely lost contact with the outside world for 6 years. Stanley presented the museum with a necklace of 37 human teeth brought back from the Congo (Figure 12.9).

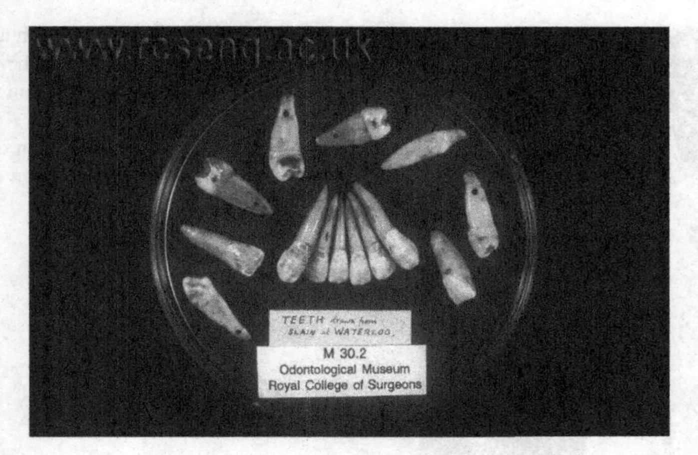

Figure 12.10 A selection of teeth removed from dead soldiers on the field of Waterloo. *Source*: Reproduced by kind permission of the Hunterian Museum at the Royal College of Surgeons.

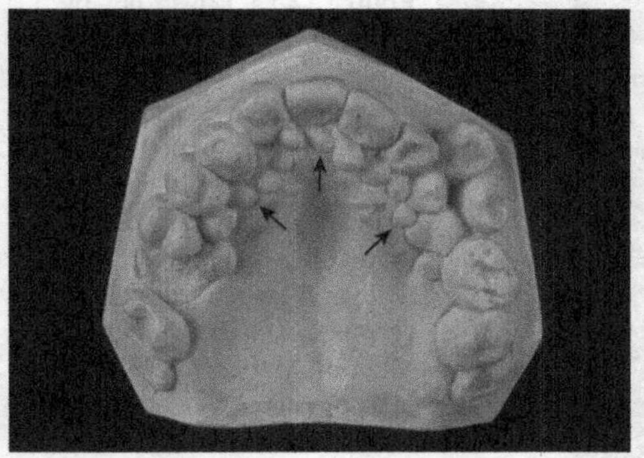

Figure 12.11 Plaster cast of upper jaw of patient containing 11 small extra teeth, some of which are arrowed. *Source*: Reproduced by kind permission of the Hunterian Museum at the Royal College of Surgeons.

In the nineteenth century, many ivory dentures incorporated real teeth extracted from living or dead people (especially casualties of war) to make them look more natural. A gruesome reminder of this practise is a collection of teeth extracted from dead soldiers on the battlefield of Waterloo in 1815 (Figure 12.10).

Numerous examples can be found showing dental anomalies and pathologies, such as the presence of extra teeth or severely displaced normal teeth. Figure 12.11 is a plaster cast of the upper teeth of an adult who has developed 11 extra teeth in the upper jaw. Figure 12.12 shows the skull of a tiger whose huge upper canine tooth is misplaced and has come to lie in the middle of the roof of the mouth, a condition also common in humans.

As a reminder of unregulated working conditions in the nineteenth century, the museum possesses examples of 'phossy jaw' (Figure 12.13). From the latter half of the nineteenth to the beginning of the twentieth century, match heads were coated with a flammable chemical containing white or yellow phosphorus. Unaware of the

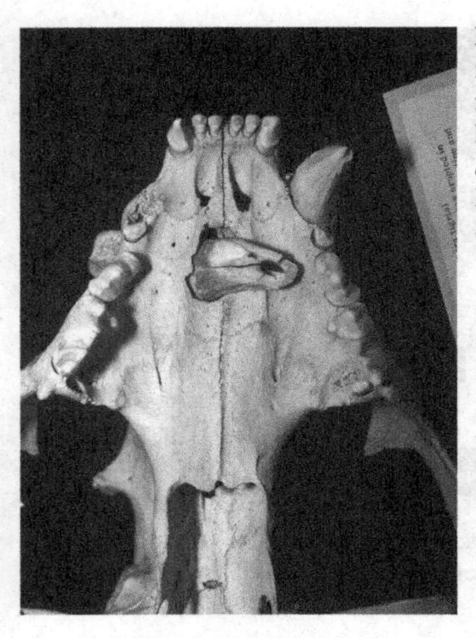

Figure 12.12 Upper jaw of a tiger with a misplaced canine tooth (tusk/fang) in the roof of its mouth. Such a condition is also common in humans.
Source: Reproduced by kind permission of the Hunterian Museum at the Royal College of Surgeons.

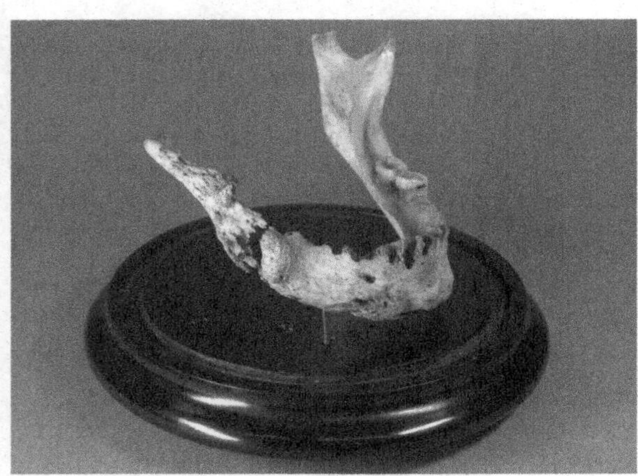

Figure 12.13 Lower jaw of matchstick factory worker suffering from phosphorus poisoning. There has been significant destruction of bone (osteonecrosis) on the right side of the jaw (compare with normal jaw in Figure 13.2).
Source: Reproduced by kind permission of the Hunterian Museum at the Royal College of Surgeons.

damaging effects of phosphorus, the poorly paid workers (mainly female) could not avoid it entering their bodies by inhalation of the fumes. The phosphorus settled in the bones, especially the lower jaw, where it caused bone death (osteonecrosis) and ulceration. This resulted in severe chronic pain, abscesses, swollen gums, tooth loss and disfigurement. The jaws would glow a greenish-white in the dark and an early death would follow. In substituting harmless red phosphorus, by the beginning of the twentieth century, this disease was eradicated.

Surprisingly, osteonecrosis still occurs today, although not in relation to matchstick making. Certain phosphorus-based drugs known as bisphosphonates are now administered as an effective treatment for cancers that have spread into bone. These drugs have occasionally produced osteonecrosis as a side effect, although the precise mechanism is not known, nor why the jaws are selectively targeted.

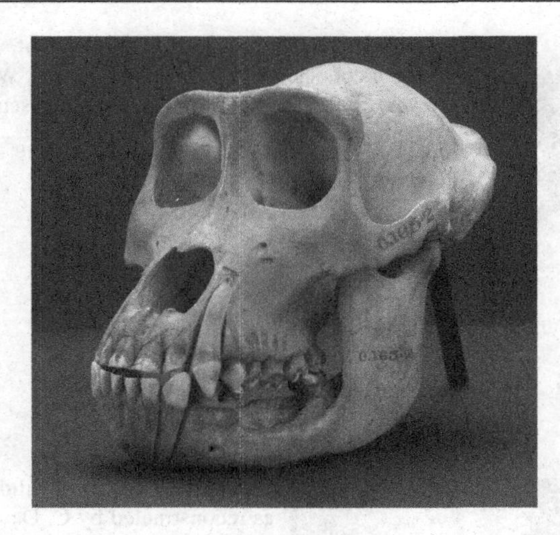

Figure 12.14 Gorilla skull with catalogue number G165.2. There is documentary evidence that Dawson drew this skull and probably used it in faking tooth wear on specimens associated with the Piltdown forgery. However, the animal has had a severe condition in the lower-left molar region with the loss of two molar teeth which has led to abnormal wear.
Source: Reproduced by kind permission of the Hunterian Museum at the Royal College of Surgeons.

One museum specimen of considerable historical importance is the skull of a gorilla, uninterestingly labelled as catalogue number G165.2 (Figure 12.14). To put this into its correct context, it is necessary to go back to December 1912. At a momentous meeting of the Geological Society of Great Britain, a joint paper was communicated by Dr Arthur S. Woodward, who specialised in fossil fish at the British Museum of Natural History, and Charles Dawson, a solicitor and amateur archaeologist.

The paper described fossil material found by Dawson in an ancient gravel pit at Piltdown in Sussex. Dawson showed some skull and jaw fragments he had discovered to Woodward, who was interested enough to return with him to continue the excavations. They subsequently found additional material, including isolated animal teeth and flint implements. Occasionally, they were helped by a Jesuit student named Pierre T. de Chardin (later to become an internationally recognised palaeontologist and theologian).

The skull fragments they found clearly belonged to a large-brained member of the human species. The lower jaw fragment contained two worn molar teeth and was ape-like (Figure 12.15). Because the fragments were lying close together, Woodward and Dawson proposed that the upper and lower jaw fragments came from the same individual. Despite having little material to work with, they reconstructed the whole of the skull and claimed it represented a new species of very early man (Figure 12.16). In honour of the discoverer, it was given the name *Eoanthropus dawsoni* (dawn man of Dawson).

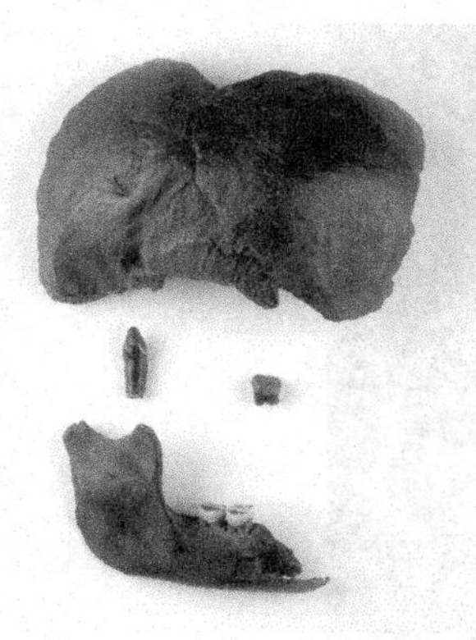

Figure 12.15 Casts of the fragments of skull bones and teeth collected by C. Dawson and A.S. Woodward.
Source: Courtesy of the Hunterian Museum at the Royal College of Surgeons.

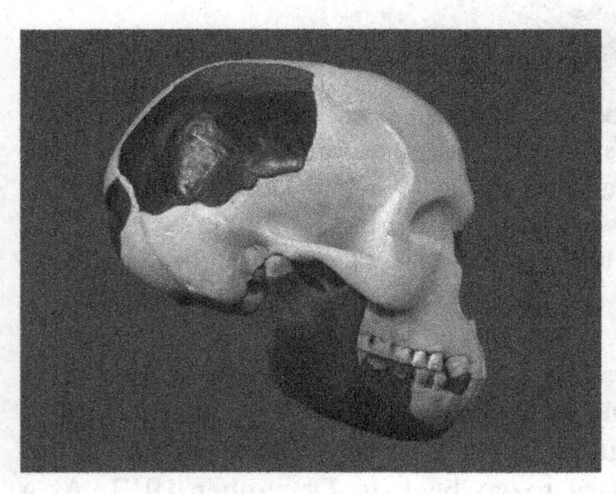

Figure 12.16 Cast of Piltdown man as reconstructed by C. Dawson and A.S. Woodward in 1912. The parts actually found are indicated in black. The rest of the skull in white has been surmised. The reconstruction depicts the upper jaw and skull with clear human features and the lower jaw with features characteristic of a great ape.
Source: ©The Natural History Museum, London.

To realise the worldwide sensation created by this discovery, it is necessary to consider the understanding of human evolution at the time, when it was thought that modern humans had evolved from primitive, ape-like ancestors. Only a few fossils had so far been discovered in Europe and the Far East, such as Neanderthal man and Java man. However, these were relatively recent ancestors already possessing the main features comprising the genus *Homo*, particularly the large brain, the human-like teeth and the ability to walk upright.

The hunt was on for an earlier, intermediate ancestor or 'missing link', which was expected to show features that were part human and part ape. As presented at the Meeting in 1912, Piltdown man seemed to fulfil all these criteria as the skull was decidedly human and the lower jaw distinctly ape-like. The site where the remains were found, together with their stained appearance, implied that the fossil was considerably older than anything previously found. Above all, it was discovered in England, indicating that early man had evolved in 'God's great country' and not in Africa. Piltdown man became a media sensation.

From the outset, not all scientists agreed that the upper and lower jaws belonged together, taking the view that the upper jaw was from a human but the lower jaw came from an ape, such as an orangutan. This would have been easy to confirm had some other parts of the lower jaw been present, such as the head of the mandible, which helps form part of the jaw joint, the midline of the mandible where it forms the chin and more teeth. Unfortunately (fortuitously?), these parts were absent. In subsequent excavations, additional small pieces of a human skull were found close together with a worn, ape-like, canine tooth, convincing some sceptics that the original reconstruction was correct and that there was no doubt that the fragments had belonged to the same fossil.

For the next 40 years Piltdown man held an important place in the expanding field of human evolution, being supported by major English scientists such as Sir Arthur Keith, Conservator of the Hunterian Museum at the Royal College of Surgeons of England.

Trouble began when South African anatomist Raymond Dart wrote in 1925 of his discovery of an almost complete fossil skull of a young individual in Taung, in the North West Province of South Africa (see also Chapter 13). He claimed it was ancestral to man, naming this fossil *Australopithecus africanus* (southern ape of Africa). The skull was small and ape-like, but the teeth were more human – the opposite to Piltdown man. Clearly Piltdown man and the Taung skull could not both be ancestral to modern man. The powerful support for Piltdown man, especially from Sir Arthur Keith, ensured that initially the Taung skull did not achieve the importance it subsequently attained in the history of human evolution. Although no further fossil material similar to Piltdown man was ever found, despite years of effort, other discoveries threw doubt on its authenticity and on its reconstruction.

Although scientists had access to the Piltdown man remains, no one bothered to undertake a detailed study of it. When finally undertaken by Drs W.S. Weiner, K.P. Oakley and W.E. Le Gross Clark and published in 1953, the results revealed that Piltdown man was a forgery. Using techniques such as measuring fluorine levels to help age the bones, it was shown that the upper jaw was human, but from the medieval period, and only about 620 years old. The fragment of the lower jaw was from an orangutan about only 500 years old. The tooth wear was not natural but had been produced artificially with an abrasive tool. Additionally, the teeth should have shown different degrees of wear. As the anterior molar erupts considerably earlier than the canine, it should have displayed more wear. In fact, the two teeth showed the same degree of wear.

The worn canine tooth of Piltdown man was found to be from a chimpanzee. It had been artificially abraded to imply substantial tooth wear and therefore to come from an old animal. However, X-ray analysis showed it had a large pulp cavity and belonged to a young specimen.

The various animal teeth found with Piltdown man were genuine fossils, but the elephant molar had originated in Tunisia and the hippopotamus tooth was from either Sicily or Malta. Finally, all the specimens had been stained artificially to make them look old and of the same age. In summary, not one piece of the material found in the Piltdown gravel pit genuinely belonged in that situation.

Following the news that Piltdown man was a gigantic hoax that any close inspection should have revealed, the question immediately arose as to the identity of the perpetrator(s). For anyone interested in whodunits, this tale has more twists and turns than the best Agatha Christie detective novel, yet this deception is no fiction. The list of those who had the opportunity, skill and motive (be it fame or vengeance on an unsuspecting colleague) grows ever longer as new information surfaces. Apart from the obvious name(s) of those who found the material are those of academics, and even technical staff, in important institutions such as the British Museum of Natural History and the Royal College of Surgeons of England. At the last count, up to 20 different people have been implicated as the forgers, with the general acceptance that at least two people would have been needed to provide the scientific and technical expertise to pull off such a stunt.

This detailed account of the Piltdown scandal has been included because it is assumed by most people that Charles Dawson was implicated. This relates to the fact that he is now known to have committed other, albeit less serious, frauds. Until recently, little hard evidence to implicate him has been forthcoming. However, through the researches of Dr Caroline Grigson, a museum curator at the Royal College of Surgeons of England, there is evidence implicating Dawson. This is in the form of a drawing of a gorilla skull that Dawson sent to Woodward. It showed tooth wear that occurred in the front teeth. This drawing was made from an actual specimen still on display at the Hunterian Museum of the Royal College of Surgeons of England, catalogue number G165.2 (Figure 12.14). Dawson was known to have visited the museum for that very purpose. He took a wax impression in the canine region of the skull, and minute traces of the wax were still evident on the teeth nearly a hundred years later. It may well be that the artificial grinding of the fossil chimpanzee tooth incorporated into the Piltdown hoax was based on Dawson's observation of this gorilla skull. Of added interest is that the wear of the front teeth present in this gorilla is probably abnormal as the animal had already lost the first two molars on the left side of the lower jaw from an unknown cause.

A Special Denture

The item with the catalogue number RCSOM/K 20.9 in the museum might not look very interesting. It is a partial upper denture containing six teeth: four front

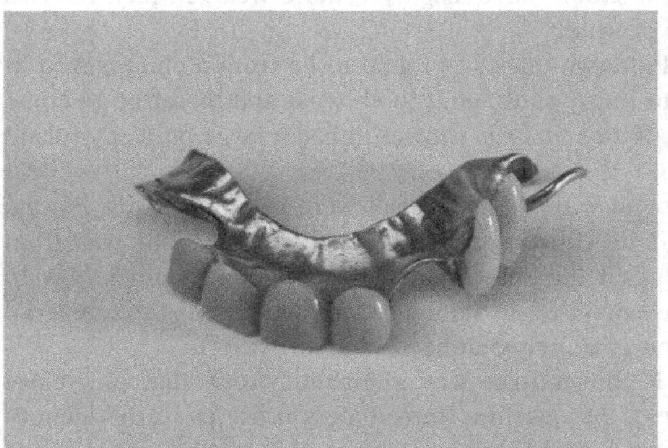

Figure 12.17 Churchill's denture.
Source: Courtesy of the Hunterian Museum at the Royal College of Surgeons.

ones and two farther back on the left side (Figure 12.17). The teeth themselves are small. There are three clues that suggest it might be special. The first is that the teeth are carried on a gold base with two clasps made of platinum, to give it extra stability. The second is that it was designed by (Sir) Wilfred Fish, one of the foremost dentists in England. The third is that the dental technician who made the dentures, Mr Derek Cudlipp, had his draft papers for the Second World War personally torn up by no less a person than the British Prime Minister, who said that Cudlipp would contribute far more to the war effort by fixing his dentures than he would by fighting on the front lines.

The denture in question is of major historical interest because it belonged to (Sir) Winston Churchill, who became Prime Minister in 1940, during the darkest days of the war, and who led Britain to victory in 1945.

It is not known for certain when, why or how Churchill lost his upper teeth. As public speaking was so important to his career, every time he had an important speech to give he must have been conscious of his denture and the fear of it becoming loose. Could he also have been dissatisfied with images of himself smiling and showing his teeth? Pictures of a smiling Churchill and showing his teeth do not seem to exist, and during the war he was rarely seen without a cigar in his mouth, perhaps another way of diverting attention from his need to smile. The state of his lower teeth is not known.

Churchill was an inspirational leader and orator. His radio wartime speeches were listened to avidly by the British people and were memorable both for content and in his manner of delivery. He did not use gesticulations and table-thumping but spoke in a slow, deliberate, restrained manner and with gravitas. One feature of his speech was a lisp, which meant the sibilant ‘s’ was exaggerated. Wilfred Fish designed the denture specifically to allow Churchill’s lisp to be maintained. This was achieved by having an imperfect seal at the back of the denture where it contacted the palate. Three identical dentures were always available and worn on a rotational system and repaired when necessary, particularly as Churchill was in the habit of throwing them across the room when things were not going his way. In recognition of his vital contribution to the war effort, in 1954 Fish was awarded a knighthood. In the letter to Fish informing him of this, Churchill was not above asking him to tighten up an enclosed set of dentures (Figure 12.18). One of Winston Churchill’s upper dentures was recently sold at auction for £15,200 ($24,000).

As orators, the contrast in delivery between Churchill and Adolph Hitler could not be greater. Hitler’s speeches were not only dynamic and mesmerising but also carefully orchestrated, proclaimed in theatrical settings to portray him as the deliverer of the German people. Hitler prepared and practised his speeches so they could be given without recourse to notes. He also knew the value and power of gestures to illuminate his speeches and could become very animated, with his voice often rising to a crescendo. Unlike Churchill, his teeth were frequently visible in photos and on film. Could it be that he had better teeth than Churchill?

Hitler’s dentition was little better than Churchill’s. In the upper jaw, he had only four of his own front teeth. In the lower jaw, he had 10, including all the front ones. However, unlike Churchill’s removable plate, Hitler had complex fixed

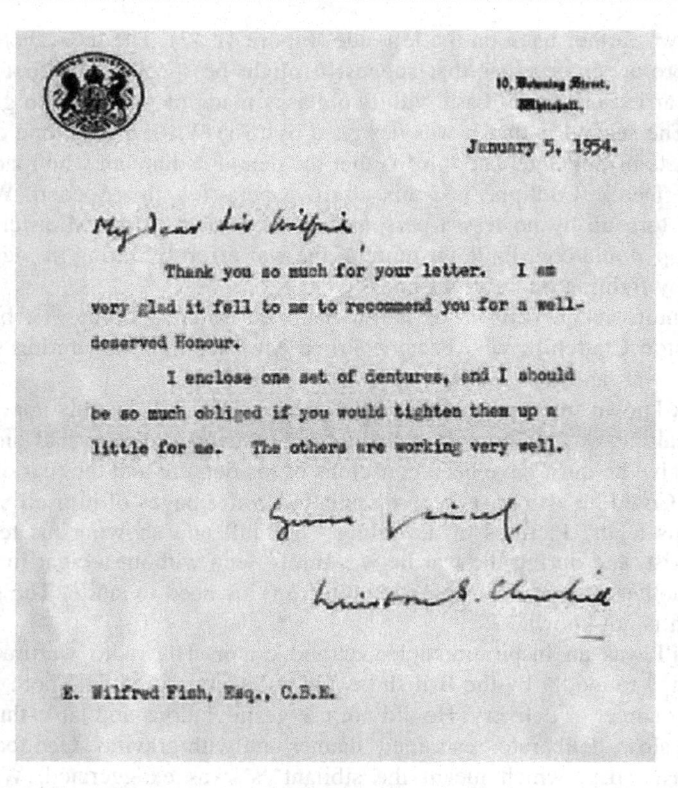
10, Downing Street,
Whitehall.

January 5, 1954.

My dear Sir Wilfrid

Thank you so much for your letter. I am very glad it fell to me to recommend you for a well-deserved Honour.

I enclose one set of dentures, and I should be so much obliged if you would tighten them up a little for me. The others are working very well.

Yours v[ery] [sincerel]y

Winston S. Churchill

E. Wilfred Fish, Esq., C.B.E.

Figure 12.18 Letter written by Sir Winston Churchill to his dentist Sir Wilfred Fish. *Source*: Published in the *British Dental Journal*, 2008, **204**:286; published online on 22 March.

bridgework undertaken to fill in the gaps (Figure 12.18). Perhaps it was because his dental bridges were firmly supported that Hitler had the confidence to be such a forceful speaker. After he committed suicide in the Fuhrer Bunker in Berlin in 1945, his body was incompletely cremated. Among the remains collected by Russian forces were his skull and teeth. It was from his complex dental bridgework that the remains were positively identified as being those of Hitler (Figure 12.19).

The author must declare a personal interest in John Hunter. It has been my privilege to hold the post of Honorary Curator of the tooth collection associated with the Hunterian Museum at the Royal College of Surgeons of England. The museum has a world famous collection of specimens related to the history of general surgery and pathology, together with a wonderful series of paintings and prints. It has been refurbished recently to provide an atmosphere deserving of the memory of John Hunter, a man of greatness. It is open and free to the general public from Tuesdays to Saturdays, so do pay us a visit. It will be well worth your while.

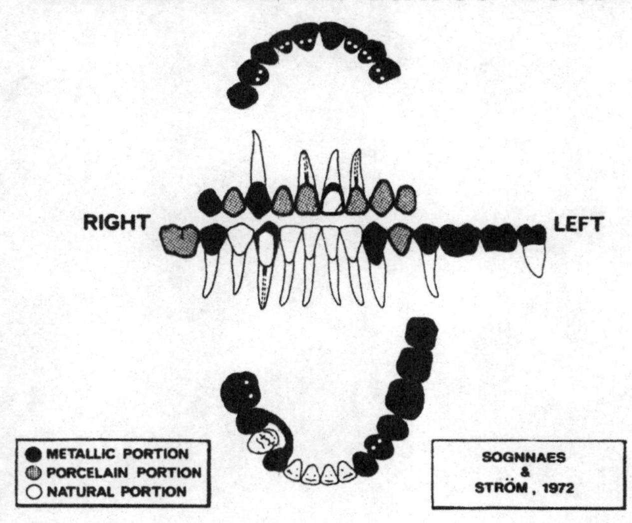

Figure 12.19 Diagram showing the features of Hitler's teeth and fixed dental bridges in both upper and lower jaws.
Source: From R.F. Sognnaes and F. Strom, 1973 and editors of *Acta Odontologica Scandinavica*.

13 Why Should It Matter How Long Our Ancestors' Teeth Took To Develop?

Soon after taking up my first teaching appointment at the University of Bristol in 1966, I decided to attend an evening lecture on the 'Evolution of Man' by Dr Louis Leakey. At the time I did not know who Leakey was. As the lecture theatre was capable of seating 400 people, I did not feel the need to get there early and turned up 5 min beforehand, anticipating ample empty seats. To my surprise, it was 'standing room only', and I was uncomfortably crouched on the floor of the aisle, alongside numerous others.

I was enthraled as slides of Olduvai Gorge in the Rift Valley of Kenya and Tanzania were shown, followed by a dramatic account of how early humans first evolved in Africa. Images of the fossilised skulls of two new species of our ancestors that Leakey and his wife, Mary, had recently unearthed created even more excitement. The more primitive and robust one, because of its large grinding teeth, had been nicknamed 'Nutcracker man' (see Figure 7.3A) and given the scientific name of *Zinjanthropus boisei* (Zinj = East African region of Zanj, anthropus = ape or ape-human, boisei = Charles Boise, who financed the expedition). It was assumed at the time that these teeth had been adapted to chew a tough diet of nuts and seeds (however, see Chapter 7, page 104). The second, less robust (gracile) fossil with a comparatively larger braincase and more human-like teeth was called *Homo habilis* (handy man). This name was given because it was thought that *Homo habilis* had made the flint tools found nearby and was thus the earliest tool-making human ancestor. These discoveries and the fact that the fossils were about 2 million years old had created headlines worldwide. The climax of the lecture came at the end, when Leakey dramatically unveiled casts of the two fossils (Figure 13.1).

Human Evolution

Humans belong to a zoological family called the Hominidae that includes the living great apes (orangutans, chimpanzees and gorillas). Within this family, there is a subgroup that habitually walked on two feet (bipedal). Modern humans as well as all the fossil species that were bipedal (such as *Homo erectus*, *Australopithecus* and even earlier species) belong to this group. They are referred to collectively as hominins.

Nothing but the Tooth. DOI: http://dx.doi.org/10.1016/B978-0-12-397190-6.00013-4

Figure 13.1 Dr Louis Leaky examining the skull of *Zinjanthropus boisei*. *Source*: http://en.wikipedia.org/wiki/File:Louis_Leakey.jpg. This image is a work of a Bureau of Land Management* employee, taken or made during the course of an employee's official duties. As a work of the U.S. federal government, the image is in the public domain.

Up to the time of Leakey's discoveries, few early fossil hominins had been found, and those that had were mainly recent (up to 500,000 years old) and very close relatives of modern humans, such as Neanderthals in Europe and *Homo erectus* in China and Java. Up to 1953, the specimens also included the remains of Piltdown man (which is discussed further in Chapter 12, page 167). The specimens that Leakey had unearthed were much older and had more ape-like features. Subsequently, Leakey, his wife, Mary, and their family (particularly their son Richard and daughter-in-law Maeve Leakey) were, and some still are, discovering a whole series of very early fossils extending back over 4 million years. Such discoveries have revolutionised our understanding of the evolution of modern humans.

What was apparent to me from the lecture was just how necessary a detailed knowledge of teeth was to the study of human evolution. Being composed of the hardest biological tissues, teeth are the structures most likely to be preserved in the fossil record (see also Chapter 7). Indeed, some hominin fossils are only known through isolated teeth. Not only is the general shape of teeth important but also the detailed structure of the enamel has provided information pertinent to evolution as it may have characteristics related to thickness and structure that are specific to a particular species. For these reasons, members of the dental profession have made major contributions to the study of the evolution of man.

Research indicates that modern humans and living great apes evolved from a common ancestor in Africa. The orangutan split off first around 12 million years ago, while the gorilla and chimpanzee split off later, about 10 and 7 million years ago respectively. The closeness of man to the chimpanzee is reflected in the fact that their genetic material only differs by 2%. The major differences between the living great apes and modern humans are the result of them having evolved separately for millions of years. As we go back to about 6 or 7 million years ago, the common ancestor to chimpanzees and the human line would have had primitive features from which both groups evolved. Such an ancestor would have had ape-like body proportions, ape-like teeth with prominent canines and a small brain and would not have walked upright. Over the next 6 million years in the line leading to modern humans, the dentition became more human with smaller canines, the gait became bipedal (habitually walking on two legs instead of on all fours) and the brain enlarged.

At one time, it was thought that there was a straight line of evolution leading to man with only a few intermediate forms. According to this view, *Australopithecus africanus* evolved into *Homo habilis*, which evolved into *Homo erectus*, which evolved into *Homo sapiens* (modern man). However, recent discoveries have shown that many more species existed and the evolutionary pathway is far more complicated with numerous side branches. As an example, the remains of a human species with a very small body and small brain have been discovered on the island of Flores in Indonesia and named *Homo floresiensis*. Just over 1 m tall and with a brain capacity of only 400 cc (compared with about 1300 cc for a modern human), this species lived as recently as 12,000 years ago. Nicknamed 'the hobbit' owing to its diminutive size, its relationship to modern humans is unclear. Some scientists believe that its reduced size was the result of a genetic disorder spreading within an isolated population. Others believe that it was related to its isolation where reduced food resources would favour the evolution of smaller individuals (as for the mini-mammoth, see Chapter 2, page 18).

It is difficult to decide which of the numerous, recently discovered fossils were on the family tree that led to modern humans and which were not. Much depends on agreeing about the precise features that are considered essential in order to be ascribed to our own genus *Homo*. A number of key characteristics can be discerned from physical evidence alone, such as brain size, the shape of the teeth and the ability to walk upright. For other characteristics such as language and communication, social organisation, religious beliefs and other rituals, it is not possible to assess this with any degree of certainty from a fossil skull alone (although some idea may be gleaned from the impression on the inside of a skull known to be related to the site in the modern brain used in processing speech).

When we compare modern humans (*Homo sapiens*) with our closest living relatives, the chimpanzees, major differences exist. Among the most important are that modern humans habitually walk upright and have a hand with an opposable thumb suited to making and using tools. In addition, modern humans have a prolonged period of dependency, when the infant is unable to survive independently. During this period, the infant is protected and can learn from its parents. Modern human skulls are less robust than those of great apes and lack the strong brow ridges and other bony crests. Our braincase is far larger (its capacity being approximately 1300 cc compared with 400 cc for the chimpanzee), our face is less prominent and we have a chin (Figure 13.2). Regarding the teeth, there are important differences in size, shape, angulation of the cusps and in the number of roots. One particularly distinguishing feature is that great apes always have large canine teeth which project beyond the level of the adjacent teeth. To accommodate the lower canine, there

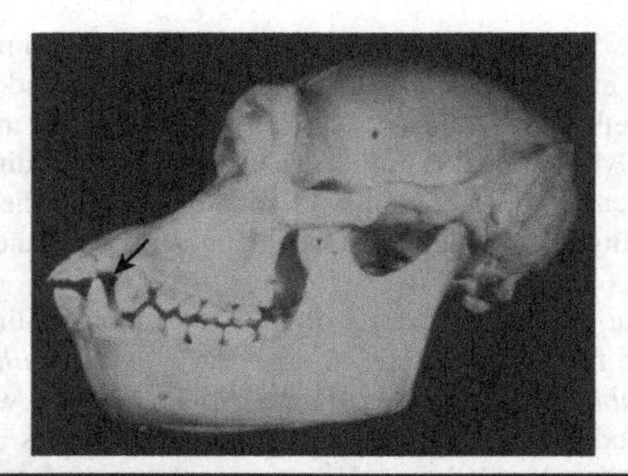

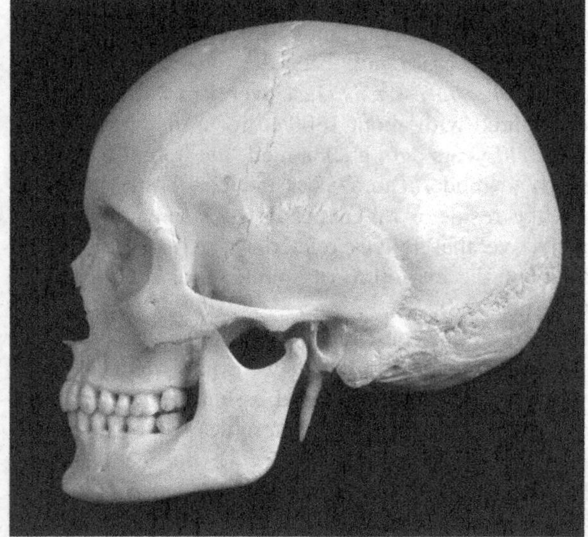

Figure 13.2 Side view of skulls of a chimpanzee (upper) and human (lower). Note the large braincase, the chin, the lack of brow ridges and the reduced size of the canine teeth in humans. Also note the space (diastema) in the upper jaw of the chimpanzee to accommodate the large lower canine (arrow).

is a gap (diastema) in front of the upper canine tooth. No such gap exists in the human upper jaw (Figure 13.2).

As an example of one of our early ancestors, consider the australopithecines. They lived 5–1.5 million years ago and constitute at least five species. Although possessing many primitive ape-like features (such as a small braincase, the body proportions of an ape and a projecting face and lower jaw), they walked upright (bipedal), albeit not as efficiently as modern humans. This facility is reflected in the shape of the pelvis. However, as fossils are rarely found with an intact pelvis, an upright posture can be inferred from a skull by the forward position of the large opening at its base called the foramen magnum (in which lies the upper part of the spinal cord that connects with the brain). In the great apes, which do not walk on two legs, the foramen magnum lies at the back of the skull.

The most remarkable evidence of an upright posture among the australopithecines is the preservation of bipedal footprints made originally in damp volcanic ash at Laetoli, Tanzania, and dating to about 3.5 million years ago.

From the skeletal remains of australopithecines can be answered the question 'What came first – a large brain or an upright posture?' The answer is an upright posture, as brain capacity in the early, bipedal australopithecines was about 450 cc, which was not much larger than that of living great apes. It was another 2–3 million years after the first appearance of bipedalism that the brain of early fossil hominids started to increase significantly (relative to body size).

Childhood Dependency

A critical feature in the evolution of modern humans has been the development of a large brain. This development is very expensive in terms of energy requirements and therefore food intake. It takes a prolonged period of time to develop a large brain and the extra skills that become possible with it. All of this can only be achieved with greater dependency of an infant on its parents. For most animals, this extra cost in parental energy and attention is simply not worth it as their lifespan is relatively short. But when mortality rates are low and animals live longer, as in the case of humans, dependency pays off because the offspring are better equipped to survive and produce more offspring of their own. During this time of dependency and protection, the infant can learn the ways of the family group and get a head start in its fight for survival. We call this period of increased dependency 'childhood', and it seems to correlate roughly with the period between the eruption of the last deciduous molar teeth at 3 years and the eruption of the first permanent molar teeth at 6 years (see Chapter 4). Moreover, brains get bigger as childhood lengthens, and the time at which the first permanent molars erupt correlates with the time brains are almost fully grown.

In our closest living relative, the chimpanzee, the first permanent molars erupt at about three-and-a-half years of age with little or no period for childhood development. It follows that if we can discover the age at which the first permanent molars erupted in extinct hominins, we can determine whether they were more like modern humans or more like living great apes. This would help identify the first fossil groups to show an extended period of childhood.

Taung Child

A very important fossil in human evolutionary history is the Taung child, so-called because it was discovered near the town of Taung in South Africa in 1924. It consisted of an almost complete skull and was named *Australopithecus africanus* (Southern apeman of Africa) by its discoverer, Dr Raymond Dart. It was estimated to be about 2 million years old (Figure 13.3). Although initially regarded by European anthropologists as being too ape-like and so of little importance, it was eventually recognised as a fossil of major significance. The Taung child possessed ape-like teeth, but its canines were smaller and its slightly enlarged braincase had

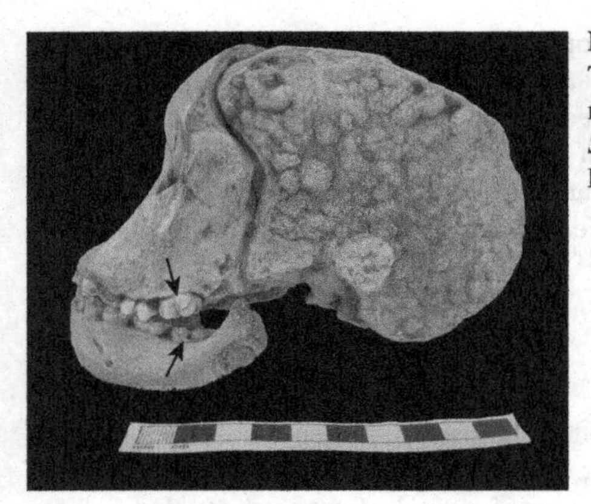

Figure 13.3 Side view of a replica of the Taung child. The erupted first permanent molars are arrowed.
Source: Courtesy of Professor MC Dean. Photographed by M Farrell.

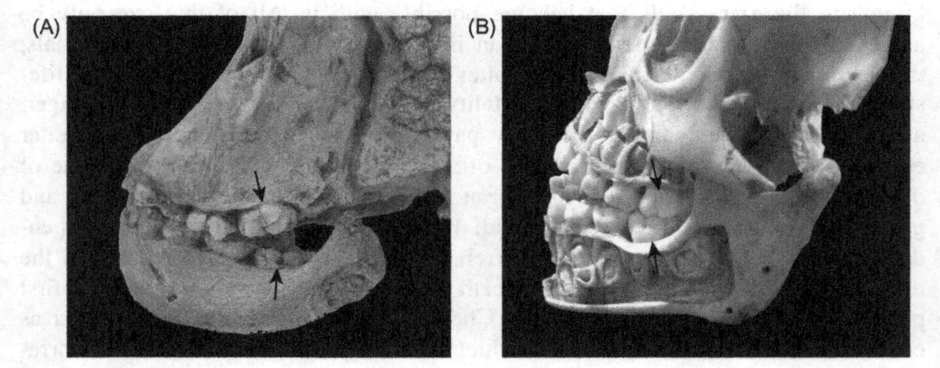

Figure 13.4 Comparison of the dentitions from a replica of the Taung child (A) and that of a 6- to 7-year-old modern human (B). Both are at the same stage of dental development with the first permanent molars erupted at the back (arrows).
Source: (A) Courtesy of Professor MC Dean. (B) Courtesy of the Hunterian Museum at the Royal College of Surgeons. Photographed by M Farrell.

some advanced features compared with an ape. For these reasons, Dart believed it to be an early human ancestor.

All the milk teeth are present in the Taung skull, but the first permanent molars had also recently erupted (Figure 13.4A). A crucial question that needs to be answered is 'How old is the Taung child?' A human child with the first permanent molars similarly erupted would be about 6 or 7 years old (Figure 13.4B; see also Chapter 4). A great ape with this type of dentition would be considerably younger, say 3 or 4 years old. This difference is quite profound as it reflects the fundamental characteristics that differentiate the two groups, namely parental dependency. If the Taung child was nearer to 6 or 7 years old, it would indicate that it had a childhood similar to humans, but if it was nearer to 3 or 4 years old, it would be more like an ape, with but a very brief childhood. When the Taung child was discovered, it was regarded as having characteristics more human than ape-like and was considered to be about 6 years of age.

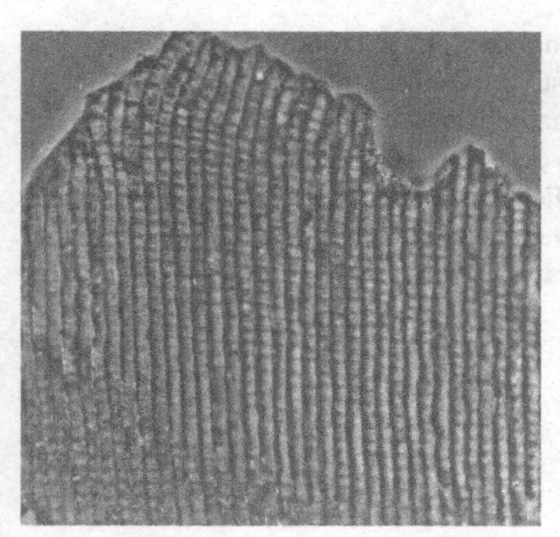

Figure 13.5 High-power view of a longitudinal section of enamel. The vertical lines represent the enamel rods, each one being divided by numerous, short, horizontal lines representing the daily incremental lines, the cross-striations. This gives each rod the appearance of a ladder. The distance between each cross-striation is about 4 μm.

Tooth Structure and Age Assessment

It is relatively easy to tell how old any living animal is by looking at its dentition and comparing it with dentitions of known age. For any extinct animal, one does not have access to specimens of definite known age. Yet knowledge of the time of eruption of the first permanent molar is critical, as it coincides in humans with prolonged parental dependency and indicates approximately when brain growth reaches its maximum size.

Is there any feature in a child's skull that will indicate its age? The answer, fortunately, is yes. Like trees, teeth have visible growth lines within them. These represent both daily (short period lines) and weekly (long period lines) periods and allow us to assess the age of an individual. In order to understand this, however, it is necessary to know something about the microscopic structure of the teeth.

Enamel is the hardest tissue in the body and covers the underlying, less mineralised dentine in the crown of a tooth (see Figure 4.2). Similar in consistency to a rock, 96% of enamel is composed of mineral (in the form of minute crystals of calcium phosphate), with only about 1% of soft protein and the remaining 3% being water. When a thin slice of enamel is looked at under the microscope, unlike a rock, it has a very ordered and characteristic appearance that shows a repeatable pattern of lines called enamel rods or prisms (see Figure 4.2C and Figure 13.5). These enamel rods run from the junction with dentine right through enamel to reach the surface. The rods are only 5 microns (μm) wide (each micron is one thousandth of a millimetre), so there are huge numbers in a tooth. The rods reflect sudden changes in the orientation of the enamel crystals that strengthen the enamel. A rock would never have such a complex repeatable structure, and this is because enamel is formed by living cells.

Cross-Striations

When enamel is forming, it comes under the influence of the 24-h daily cycle in the body, the diurnal rhythm. This cycle produces subtle changes in the enamel

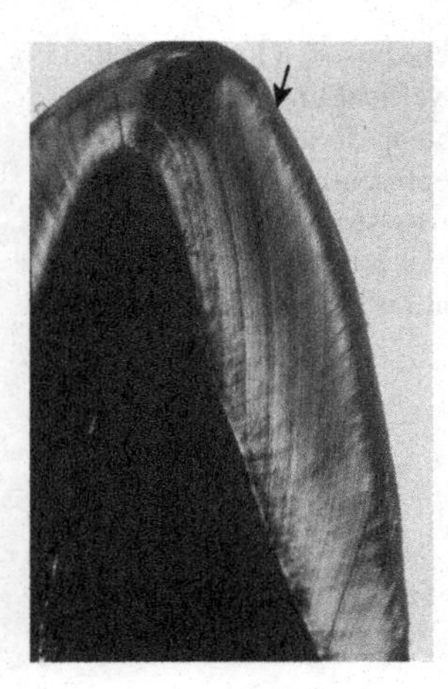

Figure 13.6 Longitudinal section of a tooth showing distribution of enamel striae (the lines passing almost vertically upwards towards the enamel surface). The first approximate 25 over the tip of the cusp do not reach the surface, but those at the sides do. Arrow indicates the site where striae first reach the surface. The striae represent weekly growth lines. *Source*: Courtesy of Drs R.J. Hillier and G.T. Craig.

structure that result in a visible line formed every day that runs across the enamel rods like the steps of a ladder. This line is called a cross-striation (short period line) (Figure 13.5) and is present about every 4 μm along an enamel rod, reflecting the amount of enamel that is formed every day. The total number of these lines that can be counted through the full thickness of enamel from tip to base will give the total number of days that the enamel of a crown took to form. The reliability of this method of aging has been confirmed by correlating the total number of cross-striations from a tooth in a child of known age.

Enamel Striae

Counting the number of cross-striations is a very laborious and difficult technique to master. It also requires considerable destruction of the tooth to prepare a suitable section to study. Clearly, this could not be used routinely on rare fossil teeth. Fortunately, there is an easier method of aging that utilises a second set of lines in enamel, the long period lines. These lines run oblique to the enamel rods and are readily seen. They are called enamel striae (see Figure 4.2C and Figure 13.6). Enamel is formed in increments, starting over the cusps of the teeth, which are deposited one on top of the other like layers in an onion. The layers are separated from each other by lines formed at weekly intervals as the enamel-forming cells move outwards. When the full thickness of enamel is formed over a cusp, the enamel striae finally reach the surface and then continue to grow downwards towards the neck of the tooth (Figures 13.6 and 13.7). The time period between two adjacent enamel striae is approximately 1 week and is confirmed by the presence on average of seven daily cross-striations between two adjacent enamel striae (Figure 13.8). The average weekly amount of enamel lying between two adjacent enamel striae is about 30 μm (i.e. 4 μm/day × 7 days). While a lot is known about the biology of daily rhythms, nothing is known about this weekly rhythm.

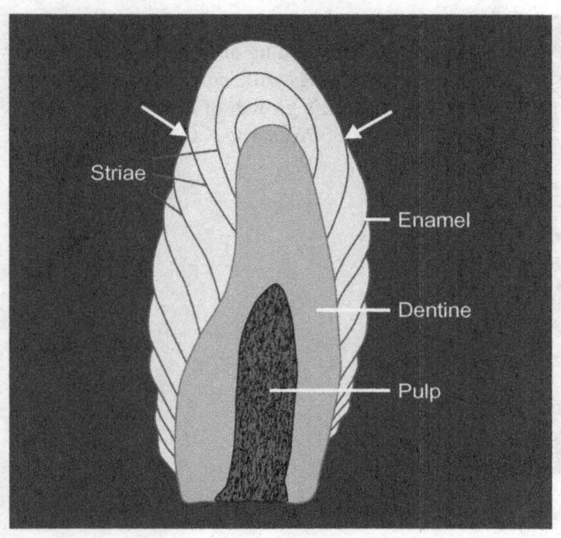

Figure 13.7 Diagram showing the distribution of the enamel striae. Only a few striae have been drawn. The first approximate 25 striae are laid down from the inside outwards, on top of each other and do not reach the surface. The arrow indicates where the striae first reach the surface and form grooves known as perikymata, which can be visualised along the rest of the side of the tooth.

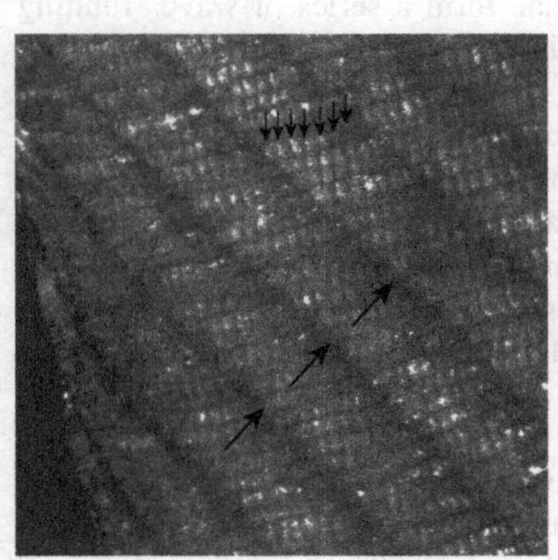

Figure 13.8 High-power view of enamel showing oblique running enamel striae, long period (weekly) incremental lines. Between each stria (large arrows) can be seen smaller vertical lines representing the short period (daily) incremental lines (small arrowheads). There are usually seven cross-striations between the striae confirming that they represent one week's growth of enamel.
Source: Courtesy of Dr D. Beynon.

Counting the number of weekly striae is a much easier method than the more laborious technique of counting all the daily cross-striations. The total number of striae across the full thickness of enamel from the tip to the neck of the tooth will give the time taken for the enamel of the crown to form in weeks.

Measuring the number of striae in thin sections of enamel is still a destructive process and could not routinely be used for rare fossil teeth. However, the striae have one other important feature that overcomes this problem. The first layers of enamel are laid down one on top of another at the tip of the cusp. It takes about 6 months before enamel is sufficiently thick to enable the striae to reach the surface at the side of the cusp (calculated by the presence of about 25 layers of striae over the cusp (Figures 13.6 and 13.7). From then on, the striae all reach the surface one after the other as they extend downwards until the crown is complete at the neck of

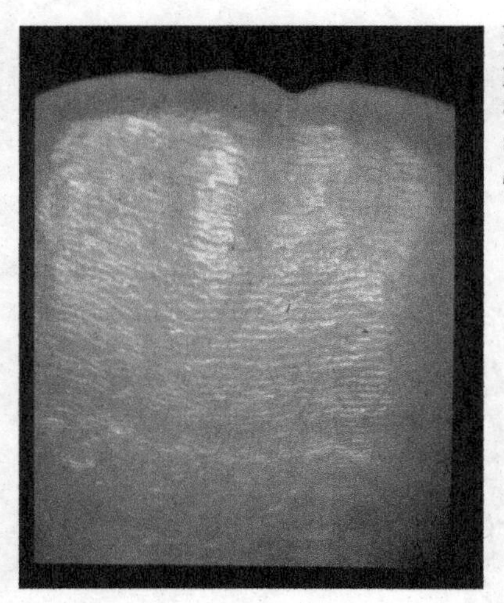

Figure 13.9 Surface of a tooth showing the perikymata grooves running around the surface of an incisor tooth in rings. Adjacent perikymata are separated by 1 week's growth of enamel.
Source: Courtesy of Professor A.G.S. Lumsden.

the tooth. Where they reach the surface, the striae form a series of waves running around the tooth and separated from each other by grooves. Each is called a perikyma (pleural = perikymata) (Figure 13.9). These grooves are usually readily visible on the surface of the tooth and can be counted directly through a microscope without any damage to the tooth. As each perikyma represents a week's growth, one can simply count the total number of perikymata present along the side of the tooth surface from top to bottom and add on about 6 months for the initial 25 or so internal perikymata that do not reach the surface. Using this non-destructive technique, a very accurate estimate of the time taken for the crown of a tooth to form can be obtained. No other method is yet available to provide such important data from skeletal remains.

Instead of counting the number of perikymata directly from the tooth using a microscope, one can also obtain an accurate impression of the tooth surface with high-quality rubber or silicone-based material. A cast can then be prepared to provide a perfect replica of the surface and the perikymata counted at leisure (Figure 13.10). The casts can be sent around the world for others to study.

Root Development

Once the crown of a tooth has fully formed, the next stage is the development of a root (Figure 11.2, Stage 5). The root is formed primarily of dentine. Structural lines are present in root dentine similar to those in enamel. About one-half to two-thirds of the root length has developed by the time the tooth first erupts into the mouth, and the root is completed 2–3 years after the tooth first erupts. The period associated with root development can be assessed and added on to that derived for crown development to get a complete figure for the time taken to form the whole tooth (crown plus root).

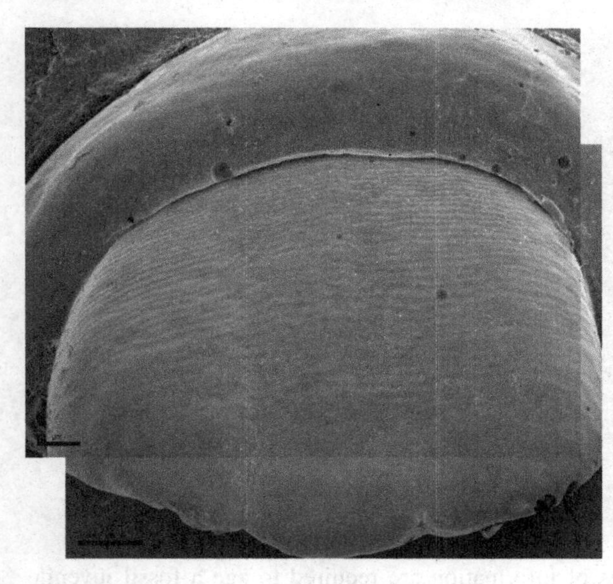

Figure 13.10 A replica of the surface of a modern tooth taken from a rubber base impression of the surface, which causes no damage to the tooth. Note the lines (perikymata) running in rings around the tooth and representing enamel striae reaching the surface.
Source: Courtesy of Professor C. Dean.

Neonatal Line

Having now obtained a figure for the time, a tooth takes to develop and erupt into the mouth of a juvenile, the final piece of information that is needed to fully age the individual is to know when the tooth first started to develop its enamel in the jaws. Again, some teeth, unlike any other surviving parts of the skeleton, have a structural feature that pinpoints the time of birth. This is because, at birth, there is a dramatic change in the blood circulation. The placental circulation from the mother that has maintained the baby is suddenly switched off and the baby has to support itself entirely with its own circulation. For the first few days immediately following birth, this change in metabolism produces a slightly different type of enamel (and dentine) that can be clearly identified in a section of a tooth and is known as the neonatal (birth) line (Figure 13.11). This neonatal line will be present in all teeth that are mineralising at the time of birth, which includes all the milk teeth and one of the cusps of the first permanent molar. The enamel formed before birth lies inside the line, and the enamel formed after birth is outside the line (Figure 13.11).

The neonatal line is so distinctive a feature that it is used in forensic science. If the skull of a baby is discovered, it may be important to determine whether the infant was stillborn or whether it had lived for a short period before its demise. In such cases, the presence of a neonatal line in the teeth would be evidence that the infant had lived for a time after birth.

For the rest of the permanent teeth, mineralisation commences after birth, so they will not possess a neonatal line. However, estimates as to when they commence to mineralise can be reasonably surmised. As an example, human permanent first incisors commence mineralisation about 4 months after birth, whereas second permanent molars commence mineralisation at about 3 years of age.

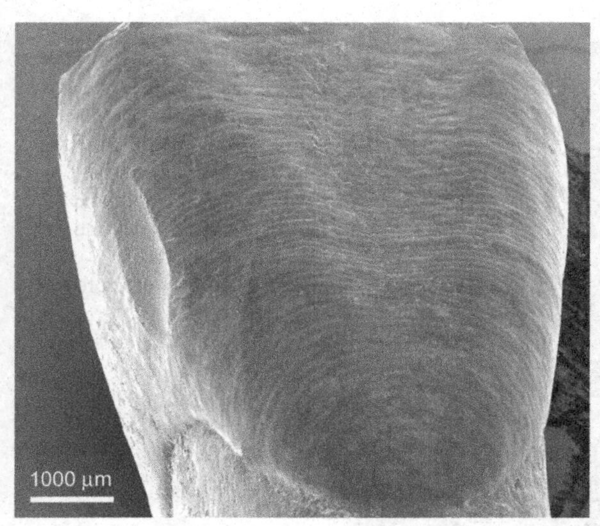

Figure 13.12 Perikymata seen as grooves around the tooth on a resin replica cast from an impression taken from the lower incisor of *Paranthropus robustus*, an australopithecine fossil approximately 2 million years old.
Source: Courtesy of Professor C. Dean.

To summarise, three pieces of information are required to age a fossil juvenile skull. It is necessary to know when the tooth commenced to mineralise, how long the crown took to form (by counting the incremental lines in enamel) and if any root is present, how long this took to reach its acquired length.

Aging the Taung Skull

Returning to the basic question of aging the skulls of our extinct human ancestors, the hominins, all the same structural features required for estimating the age of an individual using modern teeth (i.e. cross-striations, enamel striae and perikymata) are usually well preserved and can readily be seen in fossils that are millions of years old (Figure 13.12). For some, it has even been possible to prepare sections where duplicate teeth are available. Also, some teeth have fractured naturally, revealing incremental lines. The development of new, advanced microscopes has allowed scientists to obtain clear images of the enamel beneath the surface without disturbing the tooth.

Detailed analysis using all the incremental growth markers in the teeth of the Taung skull (Figures 13.3), where the first permanent molars had erupted, indicates that its age at death was 3–4 years and not, as previously stated, about 6 years. This is little different from what one would expect in a modern living great ape and contrasts with 6–7 years old for an equivalent human at the same stage of dental development. This early australopithecine, therefore, showed no evidence of delayed dental development and a prolonged period of childhood that typifies humans. A similar analysis has been undertaken for many other important extinct hominin fossil groups existing 2–4 million years ago, including representatives of Leakey's *Homo habilis* mentioned earlier (this species now being considered by many as an australopithecine and not a member of *Homo*). All the results indicate that these early

Figure 13.11 A deciduous tooth showing a neonatal line (arrow) in the enamel, separating the enamel formed before (inside the line) and after birth (outside the line).
Source: Courtesy of Drs R.J. Hillier and G.T. Craig.

hominins had an ape-like, rapid period of development, with eruption of the first permanent molars occurring early in life compared with modern humans.

Aging the Nariokotome Boy

Similar studies on growth processes in teeth have been made on our more recent ancestor, *Homo erectus*. *Homo erectus* evolved in Africa about 1.8 million years ago and subsequently migrated to Asia and then to Europe. In competition with more modern forms, it became extinct about 500,000 years ago. *Homo erectus* was the first modern human, with a cranial capacity approaching 1000 cc, and was a user of fire.

One of the most important *Homo erectus* specimens is the almost complete skeleton of a boy, dated at 1.5 million years old, at Nariokotome, by Lake Turkana in Kenya. The very rare finding of an almost complete skeleton meant that his height and probable weight could be firmly established. In addition, because many bones initially develop as a number of separate pieces that fuse together within a specific time frame, a good estimate of age could be determined from the skeleton. Results indicated that the Nariokotome boy was 160 cm tall (5 ft, 3 in) with a skeletal age of about 12 years. His dentition was remarkably complete. All his permanent teeth had erupted, except for his third molars and upper canines, and comparisons with a modern human dentition would give a dental age of about 12 years (Figure 13.13).

If aged 12 years, this would suggest he had a reasonable period of childhood, putting him close to the ancestral tree of *Homo sapiens*. It was estimated that the Nariokotome boy would have had an adult height of over 6 ft, which would make this species very tall. However, from a detailed study using incremental lines in the teeth, the age at death was estimated to be about 8 years, far lower than the first estimate made by analysing bone development. This would place the growth

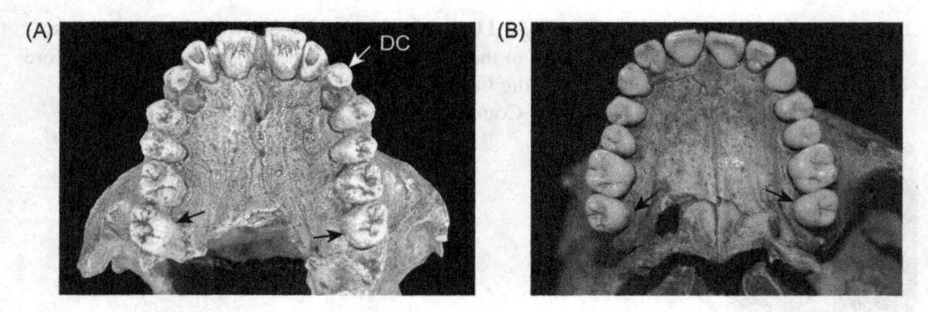

Figure 13.13 Comparison of the upper dentition of the Nariokotome boy (A) with that of a 12- to 13-year-old modern human (B). Both are at the same stage of dental development with the second permanent molars erupted at the back (arrows). The only difference is that the permanent canines have erupted in the human, while it is just erupting in the Nariokotome boy, who still has the deciduous canine (DC) in place.
Source: Courtesy of Professor C. Dean and the Hunterian Museum at Royal College of Surgeons of England. Photographed by M. Farrell.

pattern of a juvenile *Homo erectus* as being similar to that of a chimpanzee and not a modern-day human. Therefore, even *Homo erectus* lacked the slower growth phase of the human child.

Aging Neanderthals

Our closest relatives with which we coexisted in Europe were the Neanderthals (*Homo neanderthalensis*), who became extinct about 30,000 years ago. They were very similar to us, with an equal or even larger brain. However, even they appear to have had slightly faster dental development (and a shorter childhood) than modern humans, with wisdom teeth (third molars) that erupted closer to 16 years rather than to 18 years as in most of us today. This would suggest a shorter period of childhood dependency and might have been one of the reasons why they were unable to compete with us and became extinct.

Summary

This chapter has introduced some of the important fossils in our evolutionary history over the last few million years, highlighting the significance of delayed development and childhood dependency. It has also demonstrated that knowledge of the microstructure of teeth, with their incremental markings, provides unique clues to human evolution that cannot be obtained by any other skeletal remains.

14 Two Notorious People with Dental Connections: Dr Hawley Harvey Crippen (Convicted Wife Murderer) and John Henry (Doc) Holliday (Gambler and Gunman)

The dental profession can be proud of its achievements towards improving dental health. In this context, tooth decay has been greatly reduced, particularly with the discovery of the benefits of water fluoridation. This reduction in tooth decay has eliminated much of the extreme pain previously associated with dental abscesses. Similarly, gum disease can now be prevented with good oral hygiene, abolishing halitosis, bleeding gums and unnecessary loss of teeth. In the fields of orthodontics and cosmetic dentistry, the profession has pioneered advancements in improving facial aesthetics and thus restoring the self-confidence of large numbers of patients. An improved smile (see Chapter 16) can markedly increase job prospects (although exaggerated dental features have been the signature of certain show business personalities such as Terry Thomas and Ken Dodd) and marriage prospects.

Dental specialists have made significant contributions to our understanding of archaeology (see Chapter 7) and human evolution (see Chapter 13). Aspects of dental science have been important in apprehending criminals and murderers (forensic odontology). A particular example involved one of the most sensational mass murder trials of recent years when Fred West was arrested in 1994 and charged with the murder of 12 women, including his own daughter, Charmaine. Whilst in prison, he committed suicide, but not before claiming that his wife, Rose, had killed Charmaine. The police then charged Rose with her step-daughter Charmaine's murder. The only way they could make this charge stick was to show that when Charmaine was murdered, Fred West was absent with a clear alibi.

The prosecution's case ultimately depended on dental evidence. A picture of 8-year-old Charmaine was discovered in which she was smiling, with her teeth clearly visible and, at that age, still erupting. Fortuitously, the date when the picture was taken appeared on the negative. By comparing Charmaine's skull and the slightly more advanced position of her teeth than was seen in the photograph, the forensic odontologist, Prof. David Whittaker, could estimate the time that had elapsed between when the photograph was taken and when she had been murdered.

Nothing but the Tooth. DOI: http://dx.doi.org/10.1016/B978-0-12-397190-6.00014-6

This provided a relatively firm date for the murder. It was known that during this particular period, Fred West was detained in prison serving a 6.5 month jail sentence for a vehicle offence. The jury convicted Rose West of murder.

Dentistry sometimes gets an undeservedly bad press, perhaps deriving from the old days of the barber-surgeons, who, apart from shaving you and giving you a haircut, might also do some simple surgical procedures 'on the side', such as blood-letting and tooth-pulling (all without anaesthetics, of course). Two of the most widely recognised people with dental connections are 'Dr' Crippen and 'Doc' Holliday, both, unfortunately, remembered not for the benefits they have brought to mankind but for their notoriety.

Dr Hawley Harvey Crippen (1862–1910)

Dr Crippen is known for his involvement in a sensational murder trial at the beginning of the twentieth century that made newspaper headlines around the world. The case had all the ingredients to appeal to the tabloid readership of today – murder, sex, a mutilated body and an international search for the culprit.

Dr Crippen was born in the United States, in the state of Michigan in 1862. He qualified as a homoeopath in Cleveland in 1885 hence his title Dr Crippen. He married in 1888, but his wife died in childbirth 4 years later. Crippen then married his second wife, Cora Turner, in 1892 and moved to London in 1897. He and his wife had little in common. She was outgoing, flirtatious and intent on pursuing a singing career in the Music Halls under the name of Belle Elmore, although having limited talent. She liked partying and became a popular hostess. She was the dominant figure in the marriage, bullying Crippen and embarrassing him in public. Crippen, on the other hand, was a mild-mannered, quiet, respectable-looking little man, bespeckled and with a full moustache (Figure 14.1), not fitting in with his wife's circle of friends. It was his apparently benign personality which added spice to subsequent events.

Figure 14.1 Dr Crippen.
Source: http://en.wikipedia.org/wiki/File:Dr_crippen.jpg.
This is a file from the Wikimedia Commons.

Crippen found employment mainly as a supplier of patent medicines, although also in the practise of ear, nose and throat disorders, plus some ophthalmology. In order to provide a source of extra income to satisfy his wife's upwardly mobile aspirations, he went into business with a New Zealand dentist, Gilbert Rylance, forming The Yale Tooth Specialists in 1908. Their practice was based in Albion House, New Oxford Street, coincidentally sharing an office with Munyon's Homoeopathic Remedies, his previous employers. In early 1910 he got more involved in dentistry by contributing another £200 to the business. It is unclear what his precise role was, but there is no record that he carried out any clinical work as a dentist. He presumably was a business partner who provided money to buy new equipment in return for a cut of the profits. However, when the events surfaced to which his name became attached, he was widely referred to in the press as a 'dentist'.

Cora Crippen was last seen by her friends on 31 January 1910. When they enquired about her whereabouts, Crippen told them she had returned to America and showed them a letter, purportedly written by her, to this effect. However, the handwriting in the letter was clearly not Cora's. Her friends began to worry and pressed Crippen for more information. A little later, he told them she had been taken seriously ill with double pneumonia and soon afterwards notified them of her death and subsequent cremation. Indeed, on 26 March 1910 a brief announcement of her death appeared in a stage magazine called *The Era*.

Although Cora led her own life, she and Crippen had agreed to remain together because of the stigma of divorce at the time. Immediately following Cora's disappearance, Crippen was seen in the company of his young secretary, Ethel Le Neve, who was observed wearing some of his wife's jewellery. Ethel had, in fact, been Crippen's lover for 3 years. Cora's friends, being suspicious of Crippen's explanations and behaviour, eventually reported her disappearance to the police on June 30.

After carrying out some preliminary investigations, Chief Inspector Walter Drew interviewed Crippen at the offices of The Yale Tooth Specialists. During this interview, Crippen changed his story about Cora's visit to America and her subsequent death. He stated that, following one of their many rows, she informed him she was leaving him for good and that he could make up any story he liked in order to cover her absence. Crippen assumed Belle was going back to America, probably to Chicago, to live with one of her old lovers, an ex-prize fighter named Bruce Miller. When she disappeared the next morning, Crippen said he had concocted the story about her death so as not to admit publicly that she had left him.

If Crippen had kept his nerve at this time and just continued with normal activity, his life could have turned out entirely differently. The onus would then have been on the police to search for his wife. With hundreds of similar missing persons in London alone, the search for someone in America would have been quietly dropped. Unfortunately for Crippen, this is not what transpired.

Crippen and Ethel were last seen in London on 9 July 1910. They fled to Antwerp where they boarded the steamer *SS Montrose*, travelling from London to Montreal on July 20. On board ship, Crippen posed as a clergyman, getting rid of his spectacles and moustache. Ethel was disguised as his son. Just before leaving, he still found time to write the following letter to his dental partner, Dr Rylance:

> *Dear Dr. Rylance,*
> *I now find in order to escape trouble I shall be obliged to absent myself for a time. I believe the business as it is now going on you will run all right so far as money matters are concerned. If you want to give notice you should give six months' notice in my name on September 25th, 1910. I shall write you later on more fully.*
> *With kind wishes for your success,*
> *Yours sincerely,*
> *H. H. Crippen*

On July 13, 4 days after Crippen fled London, police started to search his home for any evidence relating to his wife's disappearance. On excavating the floor of the coal cellar, from which a very unpleasant odour soon arose, they uncovered the decaying remains of a butchered, human body. These consisted only of internal organs, some pieces of flesh covered by skin, patches of hair and fragments of clothing. No bones were recovered: the head, limbs and vertebral skeleton were all missing. There was no trace of the sex organs, so it was impossible to establish whether the remains were male or female. The circumstantial evidence overwhelmingly suggested that the remains were Crippen's wife. A warrant was issued immediately for the arrest of Dr Crippen, and an international hunt began, with photographs of Crippen and Ethel Le Neve plastered over all the newspapers.

The appearance and suspicious behaviour of Crippen and Ethel, who were seen holding hands aboard ship, brought them to the attention of the captain of the *SS Montrose*, Henry George Kendall. Before leaving port, he had read in the newspapers about the search for Crippen. The *SS Montrose* was equipped with wireless telegraphy, recently invented by Marconi, and on July 22 Kendall instructed the following message to be relayed to his shipping manager:

> *Have strong suspicion that Crippen London Cellar Murderer and accomplice are amongst saloon passengers. Moustache shaved off, growing beard. Accomplice dressed as boy, voice, manner and build undoubtedly a girl.*

This message was passed on to the police, who sent Chief Inspector Drew to intercept the *SS Montrose*, racing across the Atlantic aboard the liner *SS Laurentic*, that was scheduled to arrive in Canada ahead of the *SS Montrose*. As the news had been picked up by the press, the whole world avidly followed the chase in the newspapers, with the unsuspecting Crippen and Ethel totally unaware they had been located. Chief Inspector Drew boarded the *SS Montrose* as the ship entered the Gulf of St Laurence, off Quebec, and arrested the pair with the words 'Good morning Dr Crippen'. The sensation this caused was partly due to the fact that it was the first time wireless telegraphy had been used to apprehend a criminal. The press were on hand to snap the two fugitives as they descended the gangplank of the *SS Montrose* with a police escort, and these images were flashed around the world.

Crippen and Ethel were returned to London to stand separate trials, with Crippen's commencing on October 18. At his trial, Crippen was ill-served by both his solicitor, Arthur Newton, who was severely reprimanded for unprofessional conduct during the trial, and by the performance of his defence barrister Arthur Tobin. The prosecution was well served by the top barrister, Richard Muir

The trial saw the introduction of important advances in the use of forensic techniques in situations where most of the body was missing. The main prosecution witness was Dr (later Sir) Bernard Spilsbury, now regarded as the father of modern forensics. The evidence against Crippen was mainly circumstantial, as the only remains of his wife's supposed body were some organs. The main evidence in trying to identify the body revolved around the remnants of skin. The prosecution witnesses argued that the skin came from the abdomen and that it revealed evidence of a scar. This seemingly matched with Cora Crippen's previous medical history of an operation in the same area. Additional evidence in the prosecution case was the identification of traces of a poison (hyoscine) within the body, a compound that Crippen was known to have purchased recently. A pyjama top buried with the remains was said to match one that Crippen had bought.

The defence produced their own expert witnesses who denied that there was evidence of a scar and also that the skin could not definitely be identified as coming from the abdomen. In addition, they stated that the tests carried out by the prosecution witnesses did not prove the presence of hyoscine. Crippen claimed that he had used the hyoscine in very dilute doses for his patent medicines.

After a 5-day trial, Crippen was found guilty of his wife's murder, the jury requiring less than 30 min to reach their unanimous decision. Protesting his innocence to the very end, Crippen was hanged on 23 November 1910, aged 48 years.

At Ethel Le Neve's trial, which took place on 25 October, shortly after Crippen's, she was defended by a brilliant barrister, Mr F.E. Smith. He decided not to put Ethel on the stand or provide any defence witnesses. His speech provided the only defence. He portrayed Ethel as a young innocent girl taken advantage of and led astray by an older and socially superior employer. His defence was that the prosecution could not prove beyond reasonable doubt that Ethel knew that Crippen murdered his wife. Ethel Le Neve was found not guilty. She disappeared from public view and died anonymously nearly 60 years later in 1967.

Since his death, many have doubted Crippen's guilt. His mild personality did not seem to coincide with the cold and calculating character of a murderer. Many witnesses at the trial had said what a nice person he was. Having successfully disposed of the larger and most conspicuous parts of the body (skull, limbs, rib cage and vertebral column), why would he have felt the need to bury various soft tissue components, some wrapped up in his own clothing, close to his own dining room and capable of producing such an awful stench?

The most recent and compelling evidence to prove Dr Crippen's innocence was published in 2010, using state-of-the-art forensic techniques of DNA analysis. Samples of the soft tissue remains of his 'wife' used by the prosecution as evidence to convict him were tracked down to the Royal London Hospital Archives and

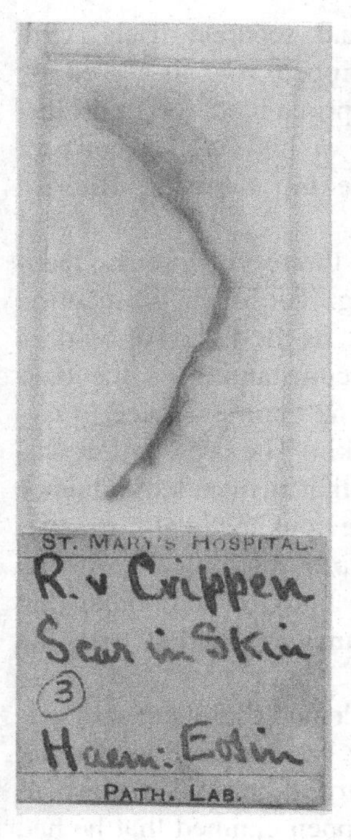

Figure 14.2 Original histology slide containing tissue from the abdominal skin taken from the body found in the basement of Dr Crippen's house and assumed by the prosecution to be that of his wife Cora.
Source: ©The Royal London Hospital Archives.

Museums. They existed on the original microscope slides prepared by Dr Spilsbury of what he believed was the abdominal skin from Cora Crippen (Figure 14.2). A team of scientists led by Prof. David Foran from the University of Michigan obtained permission to remove the soft tissue remains from one of the slides. The team carefully isolated and analysed DNA from the specimen and compared the results with three living members of Cora's family. There was no match between the microscopical material and the family members, indicating that the tissue could not possibly have come from Cora Crippen. However, the most startling result was that the skin was from a man and not a woman. Assuming the validity of these results, it is clear that Crippen was found guilty on completely flawed evidence. Furthermore, two obvious questions arise from this study: What really happened to Cora Crippen? and Who was the man buried in Dr Crippen's coal cellar? Based on this evidence, descendents of Crippen are engaged in trying to get a belated judicial pardon for him.

On Crippen's death certificate (Figure 14.3), the cause of death is written as 'fracture of vertebrae by hanging. Executed by law.' Under the heading 'Occupation', for all to see, is 'Dentist'.

Over a century later, Dr Crippen's name continues to be well known in England. A number of films, books and articles about him are in circulation. There has even been a musical about him called 'Belle or the ballard of Dr Crippen'. He is still prominently displayed as a bogeyman in the chamber of horrors at

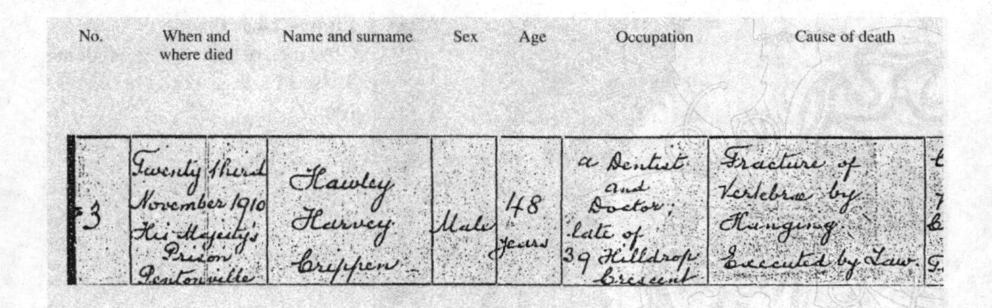

No.	When and where died	Name and surname	Sex	Age	Occupation	Cause of death
3	Twenty third November 1910 His Majesty's Prison Pentonville	Hawley Harvey Crippen	Male	48 years	a Dentist and Doctor; late of 39 Hilldrop Crescent	Fracture of Vertebrae by Hanging. Executed by Law.

Figure 14.3 Dr Crippen's death certificate, showing cause of death and his profession. *Source*: Courtesy of Home Office, UK.

Figure 14.4 Dr H. Crippen in the Chamber of Horrors at Madame Tussaud's Waxworks in London.

Madame Tussaud's Waxworks in London (Figures 14.4), where he stands behind bars with a plaque (Figure 14.5) announcing:

Dr Hawley Harvey Crippen
1862–Hanged 1910
Wife poisoner
Arrested on board ship following a
Telegram sent to Scotland Yard.

Dr Crippen's claim to be the most infamous person with dental connections is not as strong as the second contender. This candidate is known throughout a much wider region, especially in North America. He also has a stronger connection with the dental profession in that he was a qualified dentist. In addition, he became such a legendary folk hero as to be a main character in numerous Hollywood and television films. The incident with which his name is most closely related is still re-enacted on a daily basis as a major tourist attraction. His name is John Henry Holliday,

Figure 14.5 Plaque in Chamber of Horrors at Madame Tussaud's Waxworks in London.

Figure 14.6 Image purported to be of Doc Holliday taken in the early 1880s.
Source: http://en.wikipedia.org/wiki/File:Doc_Holliday.jpg. This image is in the public domain in the United States. In most cases, this means that it was first published prior to 1 January 1923. This is a candidate to be copied to Wikimedia Commons.

known more familiarly as 'Doc' Holliday. He was a major combatant in the Gunfight at the O.K. Corral (which in fact did not take place at the O.K. Corral but in a vacant lot nearby). This event linked him forever with another Western legend, Wyatt Earp.

'Doc' Holliday (1851–1887)

John Henry Holliday (Jr) (Figure 14.6) was born in the United States on 14 August 1851 in the town of Griffin in the state of Georgia. At that time, Griffin was frontier country, with people arriving to colonise the Western regions of North America. In the Southern states, the values most respected were courtesy, loyalty, toughness and a willingness to fight for one's principles. Holliday seemed to encompass many of these attributes. He had a happy childhood amongst his extended family and took readily to the outdoor life, absorbing the skills of hunting and shooting.

Holliday's happy childhood ended when the state of Georgia, with its interest in maintaining slavery, declared itself independent of the Union in 1860. Ten-year-old Holliday witnessed the excitement as the male members of his family volunteered to join the Confederate forces at the start of the American Civil War in 1861. Everyone was confident they would win, as their cause was a worthy and principled one and they felt honour-bound to defend their liberty and way of life. Holliday was left at home with his mother, Alice, and their slaves.

Following the initial successes of the Southern armies, the tide of battle changed and victory for the Union forces of the North became inevitable with their advantages of industrial might and population. When General Sherman's forces started their march through Georgia, Holliday's family sold their land, which lay in Sherman's path, and moved to the small town of Valdosta, some distance from the main conflict. After the Civil War ended in April 1865, a difficult period ensued for the people of Georgia, with its occupation by the successful forces of the North, especially with the presence of armed black soldiers. Added to this, Holliday's mother, Alice, died of consumption, now called tuberculosis, at the age of 36, in September 1866. There was no cure for this highly infectious illness, which, at the time, was said to be responsible for about one-fifth of all deaths in the region. As we shall see, this same disease was to have a profound effect on the life of her son.

The death of his mother when he was 15 years old and the loss of her restraining influence may have precipitated a gradual change in Holliday's character from then on. Coupled with this bereavement, he lost respect for his father, who remarried just 3 months later, without allowing for the customary period of mourning. Holliday eventually changed into the individual that typified those legendary figures of the 'Wild West', a tough, fearless, hard-drinking gambler and gunfighter.

At school Holliday proved to be an intelligent and capable student. Under the influence of the local dental surgeon, Dr Frink, he chose to pursue a similar career and enrolled at the Pennsylvania College of Dental Surgery in October 1870. He was 19 years old.

The Pennsylvania College of Dental Surgery was one of the foremost dental schools in the country. The course was a demanding one, and Holliday proved to be a capable student. After an initial 6-month period of study at the dental school, he returned to his hometown of Valdosta to get some work experience and training at the practice of Dr Frink, who took on the role of his supervisor. Holliday returned to the Pennsylvania College of Dental Surgery in September 1871 and completed the course in March 1872 (Figure 14.7). A student had to be age 21 to obtain a dental qualification, so to pass the next few months, Holliday travelled to St Louis, Missouri, to work in the dental practice of one of his older classmates. It is likely that here he first met Mary Katharine Harony (also known as Big Nose Kate), who had a tempestuous on/off relationship with Holliday for the rest of his life. She was a hard-drinking, tough, larger-than-life woman who, no doubt, needed

Figure 14.7 Graduation photograph of Doc. Holliday taken in 1872.
Source: http://en.wikipedia.org/wiki/File:DocHollidayCloseUp.jpg. This image is in the public domain in the United States. In most cases, this means that it was first published prior to 1 January 1923. This is a candidate to be copied to Wikimedia Commons.

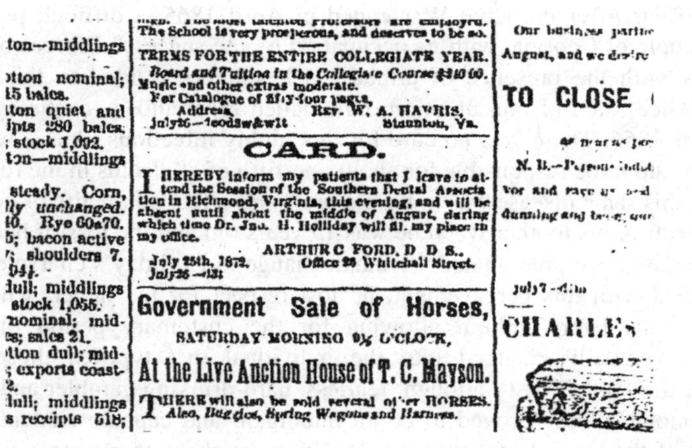

ton—middlings
otton nominal;
15 bales.
tton quiet and
ipts 280 bales;
; stock 1,002.
ton—middlings
steady. Corn,
lly unchanged.
40. Rye 60a70.
5; bacon active
; shoulders 7.
dull; middlings
stock 1,055.
nominal; mid-
s; sales 21.
otton dull; mid-
; exports coast-
2.
dull; middlings
s receipts 518;

The School is very prosperous, and deserves to be so.

TERMS FOR THE ENTIRE COLLEGIATE YEAR.
Board and Tuition in the Collegiate Course $210 00.
Music and other extras moderate.
For Catalogue of fifty-four pages,
Address, REV. W. A. HARRIS,
july26—1eod3w&w1t Staunton, Va.

CARD.

I HEREBY inform my patients that I leave to attend the Session of the Southern Dental Association in Richmond, Virginia, this evening, and will be absent until about the middle of August, during which time Dr. Jno. H. Holliday will fill my place in my office.
ARTHUR C. FORD, D. D. S.,
July 25th, 1872. Office 26 Whitehall Street.
july26—13t

Government Sale of Horses,
SATURDAY MORNING 9½ O'CLOCK,
At the Live Auction House of T. C. Mayson.
THERE will also be sold several other HORSES.
Also, Buggies, Spring Wagons and Harness.

Our business partie
August, and we desire
TO CLOSE
N. B.—
july7—
CHARLES

Figure 14.8 Advert in the *Atlanta Constitution*, July 1872. The advert reads: 'CARD, I hereby inform my patients that I leave to attend the session of the Southern Dental Association in Richmond, Virginia, this evening, and will be absent until about the middle of August, during which time Dr John H. Holliday will fill my place in my office. Arthur C. Ford D.D.S'.
Source: Courtesy US Library of Congress.

such qualities to endure in a relationship with Holliday. She also was to become a well-known figure in the history of the 'Wild West' in her own right.

Once Holliday qualified as a dentist, he acquired the nickname 'Doc'. He returned to Atlanta, Georgia, and his name first appears in a local newspaper, the *Atlanta Constitution*, advertising his services in the practice of a Dr Ford in July 1872 (Figure 14.8). Holliday soon returned to Griffin, where he opened his own dental practice. He agreed to become supervisor for his cousin, Robert

Holliday, who also attended the Pennsylvania College of Dental Surgery, in October 1873. It therefore seemed that Holliday would have a rosy future running a successful dental practice and becoming a respected member of the local community. However, such expectations suddenly changed, for in the summer of 1873, Holliday showed up in Dallas, Texas, as a partner in the dental practice of Dr John Seegar. The precise reason for this sudden move to Dallas is unknown. It may have been the result of an unrequited love affair. It could have been related to his health, for Holliday had contracted tuberculosis, possibly from his mother or from one of his patients. The symptoms of this disease, especially chronic bouts of coughing, were to plague him throughout the rest of his short life. He may have also been involved in a shooting incident from which he needed to escape.

It wasn't long before other activities started to fill his time, namely gambling (poker and faro) and the drinking associated with it. His notoriously heavy drinking may have helped him cope with his chronic cough. The first of his many brushes with authority in the lawless towns of the West occurred in Dallas in April 1874. Holliday was arrested for gambling, which was still forbidden in some towns.

Despite being short in stature (160 cm or 5 ft, 3 in tall) and physically weak, he was already gaining a reputation for being hot-headed, fearless and able and willing to defend himself with a gun or a knife, characteristics important for a gambler: he was not a man to be messed with. This aggressive stance was likely to be increased by his awareness of his limited life expectancy.

Holliday's dental skills were put to one side as he travelled around the gambling tables of the developing frontier towns, which must have been more profitable. These towns often sprang up overnight following a strike of gold or silver. People flooded in from all around in the hope of striking it rich. Such towns were filled with cattle farmers, prospectors, gamblers, rustlers (who chose to steal, rather than cultivate, cattle and horses) and ex-soldiers looking for quick opportunities to get rich. With so few women available, prostitution was widespread. After working hard obtaining money, men would be ready to spend it gambling, drinking and womanising, and there were many people only too happy to relieve them of their money, both legally and illegally. Once the main strikes had been worked, people moved on looking for the next one.

Although the US government and the local authorities had law enforcement officers in the form of sheriffs and their deputies, lawlessness was rife in frontier towns, and trying to enforce the law was not only difficult but dangerous. At the time, the term 'cowboy' often had a derogatory meaning, referring to people who didn't worry about breaking the law. Tough, independent men, especially drawn from Texas, the cowboys knew that some of their actions were illegal, but they did not regard themselves as criminals. They resented any interference by government or local agents limiting their freedom of action. They would undertake ventures such as robbery of stagecoaches or cattle rustling. The latter required cooperation from ranchers, saloons and restaurants, but with money to be made on all sides, not too many questions would be asked.

Over the next few years, Holliday travelled between towns such as Denver, Cheyenne, Wyoming, Dallas and Fort Worth. During this period, he suffered

Figure 14.9 Wyatt Earp, aged about 33 years.
Source: http://en.wikipedia.org/wiki/File:Wyatt_Earp_portrait.png. This is a file from the Wikimedia Commons. Commons is a freely licensed media file repository and This media file is in the public domain in the United States. This applies to U.S. works where the copyright has expired, often because its first publication occurred prior to 1 January 1923.

further arrests for gambling and violence. In Texas he first encountered Wyatt Earp, who was to play such an important role in his life, culminating in the gunfight at the O.K. Corral.

Wyatt Earp (Figure 14.9) was one of five brothers, the others being Virgil, Morgan, Warren and James. Although ambitious to make good and become rich and famous, this did not materialise throughout most of Wyatt's life. Tall and physically strong, he was fearless and could outfight almost anyone. He was quick to take offence and was never willing to step down or publicly lose face. This contrasted with his brother Virgil, who was more diplomatic by nature and would try to seek out an easier solution to a problem if it meant avoiding a fight. Wyatt had occupied various jobs in law enforcement and worked for Wells Fargo, protecting money by riding shotgun on stagecoaches. In addition, he gambled and had varied interests in mining and saloons. This was fairly normal even for law officers, who were expected to have other sources of income. Their positions were political, often requiring election by local people.

Doc Holliday appears to have first encountered Wyatt Earp in 1877. He may have been contacted later by him and advised that there were considerable opportunities waiting for a man of his particular talents and skills in Dodge City, Kansas. Doc Holliday arrived there around the middle of 1878. Dodge City was a major cattle centre and a fairly wild place. Its town marshal, Edward Masterson, had recently been killed while carrying out his lawful duties, and Wyatt Earp had been appointed as assistant marshal. As this bustling and wealthy town had no dentist, Holliday reverted to his original profession and advertised his services in the *Dodge City Times* as follows (Figure 14.10):

DENTISTRY

John H. Holliday, Dentist, very respectfully offers his professional services to the citizens of Dodge City and surrounding county during the summer. Office at Room No. 24 Dodge House. Where satisfaction is not given, money will be refunded.

With Holliday's known feisty reputation, one wonders how many patients would have dared declare not being satisfied and demand a refund!

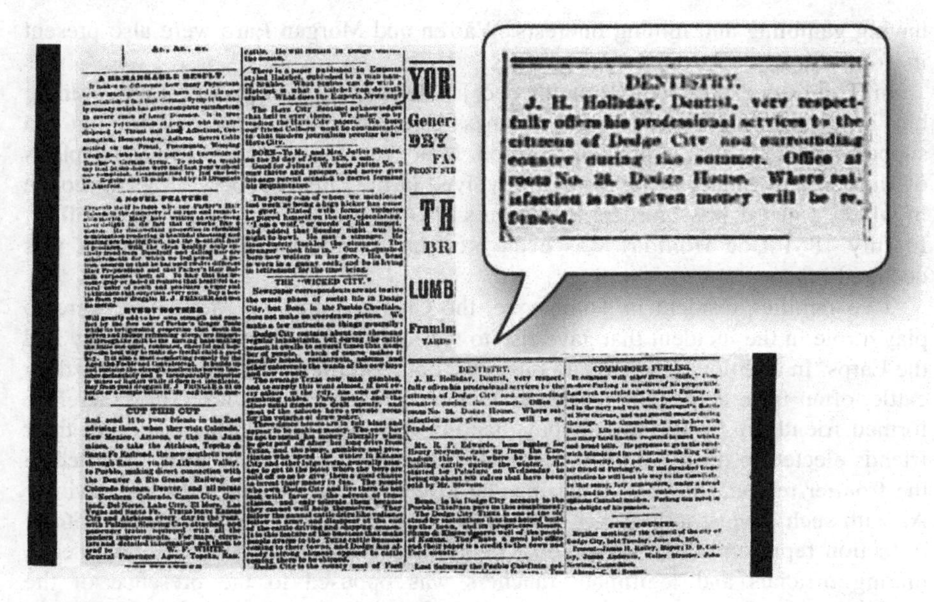

DENTISTRY.

J. H. Holliday, Dentist, very respectfully offers his professional services to the citizens of Dodge City and surrounding country during the summer. Office at room No. 24, Dodge House. Where satisfaction is not given money will be refunded.

Figure 14.10 Advert placed in *Dodge City Times* by Doc Holliday on 8 June 1878. See text for translation.
Source: Courtesy US Library of Congress.

It was during his stay in Dodge City that Holliday is thought to have saved Wyatt Earp's life by backing him up when Wyatt was surrounded by a group of cowboys intent on revenge for the killing of one of their number. This action bound the pair together for the rest of Holliday's life.

Holliday didn't remain long in Dodge City. Moving to the warmer and drier climate of Las Vegas, New Mexico, known as a recuperative site for consumptives, he became a partner in a saloon during the summer of 1879. However, further troubles with the law relating to gambling and firearms made him relinquish his saloon and move on. Following his departure from Las Vegas, Holliday never again practised dentistry, committing himself fully to the life of a gambler.

Again at the invitation of Wyatt Earp, Doc Holliday arrived in Tombstone, Arizona, in September 1880. Tombstone was a mining and cattle town, expanding rapidly following the discovery of a large silver strike. In addition to the inevitable gambling and drinking facilities, it also had the basis of a business community, with established hotels and shops.

The Earp brothers had gathered together in Tombstone, intent on seeking the fortune which had so far eluded them. Lawmen could make good money with the right political backing and with various deals on the side. Virgil had been appointed a deputy US marshal, representing the government in law enforcement, while Wyatt was a deputy sheriff representing local law enforcement but also having gambling and mining interests. Warren and Morgan Earp were also present in Tombstone.

In Tombstone, Holliday strengthened his friendship with Wyatt Earp, joining him in legal (and probably illegal) business enterprises associated with mining and supporting him with a gun when required. Many rumours abound as to the exploits of Holliday, including being actively involved in the failed robbery of a stagecoach in which a guard was murdered. Rumours also implicated the Earps in this affair. In July 1881 Doc Holliday was even brought to trial for this robbery but was declared innocent.

Two families resident in Tombstone, the Clantons and the McLaurys, were to play a role in the incident that gave rise to the legendary reputation of Holliday and the Earps. In addition to legitimate business, both families were involved in rustling cattle, often from across the border in Mexico to avoid paying tax. They therefore formed friendships with the 'cowboy' faction and used their influence to get their friends elected to office in town councils. Although the cowboys were tolerated in the frontier region, they had a detrimental effect on other legal business activities. As with such lawlessness, it was not safe to pass through the surrounding territory. A faction represented by professional people such as bankers, lawyers, engineers, mining officials and legitimate ranchers was opposed to the presence of the cowboys.

Animosity existed between the Earps on the one side and the Clantons and the McLaurys on the other. One episode fostering this distrust related to a secret deal that Wyatt Earp made with Ike Clanton in connection with the robbery of a Wells Fargo stagecoach, of which Holliday was accused but found not guilty. The plan was that if Ike Clanton informed Wyatt where to find the culprits, Wyatt would give Ike Clanton all of the considerable reward money. In return, Wyatt would get favourable publicity and the chance of being elected to a higher law enforcement position. Ike agreed to this plan, even getting an agreement that the reward would be paid should the robbers be brought in dead. In the end, this deal did not materialise as the robbers were shot dead by another party. However, the knowledge that Ike Clanton had secret meetings with Wyatt Earp, a representative of the law, and had informed on his friends would not have gone down well with the 'cowboy' faction.

This episode preyed on Ike's mind and he was very worried that Wyatt Earp would leak news of their private meeting. Indeed, during a poker game on 25 October 1881, Ike Clanton, an impulsive character at the best of times and thoroughly the worse for drink, accused Doc Holliday of knowledge of the secret meeting. Holliday resented the charge and immediately challenged Clanton to a gunfight. Ike Clanton was unarmed at the time, and Holliday told him to go and get a gun. Wyatt Earp, who was also playing in the same poker game, sent word to his brothers to come over and calm things down, which they did, threatening to arrest the two main combatants. Ike Clanton left, still talking about harming not only Holliday but also the Earps, whom he felt had taken Holliday's side in the argument. Ike Clanton continued to drink and gamble through the night.

The next morning, October 26, Ike Clanton, now armed, let everyone know he was looking for Doc Holliday and the Earps. As it was against the law to carry arms in the town, Virgil and Morgan Earp tried peacefully to disarm Clanton, but it ended up with Virgil having to club him to the ground. Instead of taking him straight into jail to cool off, Virgil arranged for Ike to be brought before a judge. While he was arranging this, he left Wyatt and Morgan Earp in charge of Ike and, no doubt, threats and counter-threats continued to be made between the parties.

An angry Wyatt Earp later encountered Ike's companion, Tom McLaury. Although Tom wasn't looking for trouble, further words were exchanged, and he was assaulted and beaten up by Wyatt. A very heated atmosphere was growing between the parties, which wasn't helped by the arrival of Tom's brother, Frank McLaury, and Ike's brother, Billy Clanton. The four of them were seen in the local gun shop buying ammunition, after which they proceeded down the street to the O.K. Corral (an enclosure for horses). They were joined by a fifth person, a friend named Billy Claibourne. Billy was a young gunfighter who liked to refer to himself as Arizona's Billy-the-Kid (the original outlaw having been killed by Sheriff Pat Garrett a couple of months earlier).

Although initially they felt no real threat from Ike Clanton, the additional presence of the three newcomers made the Earps feel their safety was in jeopardy, even though they were representing the law and would have the support of the business faction. The Clantons and McLaurys felt they had been very roughly treated by the Earps. To swallow this and appear to be intimidated would be to lose face in the eyes of onlookers and friends. Virgil, Wyatt and the newly deputised Morgan could similarly not be seen to back down. Armed, the Earps aimed to confront the Clantons and the McLaurys. If there had to be a fight, the Earps would provide one.

At about 2 pm, County Sheriff Johnny Behan, who himself had issues with the Earps, tried to defuse the situation by saying he would first attempt to get the Clantons and McLaurys to disarm. Virgil agreed that as long as they remained in the O.K. Corral, he would not go down and confront them.

At about 2:30 pm, Doc Holliday, having heard of the morning's developments, met with the Earps. Although told it was a private matter, he could not be dissuaded from joining them. No doubt the Earps realised that, if it came to a gunfight, Doc Holliday's presence could only be of benefit to them, particularly in discouraging any passing cowboys to join in. With this in mind, Virgil deputised Holliday.

In the meantime, the Clantons and McLaurys had left the O.K. Corral and were coming into the main street, where they met Sheriff Behan. He could see that Frank McLaury was armed. However, the group informed Behan that they wouldn't give any trouble, but Frank was unwilling to disarm until the Earps had also disarmed, as they had threatened violence.

It was now about 3 pm, and seeing the Earps and Doc Holliday approaching, Sheriff Behan passed the message back, thinking that it might be enough to deter the Earps and Holliday from proceeding. He may have mistakenly given them the impression that the Clantons and McLaurys were not armed. However, the Earp group brushed past him and continued on.

The image of the three Earp brothers, all about 6 ft tall and moustached, walking down Freemont Street towards the O.K. Corral to confront the Clantons and McLaurys, with Doc Holliday a few paces behind carrying a shotgun under his coat, is a climactic scene in many Hollywood films of the incident: In the film *Gunfight at the O.K. Corral*, Kirk Douglas played Doc Holliday and Burt Lancaster portrayed Wyatt Earp. In the film *Tombstone*, it was Val Kilmer as Doc Holliday and Kurt Russell as Wyatt Earp. In *Wyatt Earp*, Dennis Quaid played Doc Holliday and Kevin Costner was Wyatt Earp. In *My Darling Clementine*, it was Victor Mature who played Doc Holliday with Henry Fonda as Wyatt Earp.

The opposing groups confronted each other, not in the O.K. Corral but at a nearby vacant lot. They were separated by just a few feet: it was well known that pistols were relatively inaccurate at distance. Having understood them to be unarmed, the Earps might have been taken aback by the sight of Billy Clanton and Frank McLaury carrying guns and with rifles on the saddles of their horses nearby. Virgil Earp announced that he was going to disarm them and that they should raise their hands.

The precise order of events that followed is not known for certain. There was movement of some of the cowboys' hands, but whether they were about to disarm or were readying to fire cannot be determined. Wyatt Earp and Frank McLaury may have been the first to shoot, more or less simultaneously. Whatever the cause, after a brief exchange of bullets lasting about 30 s, Frank McLaury, Tom McLaury and Billy Clanton lay dead or dying. Tom McLaury had been killed by Doc Holliday's shotgun. Only Wyatt Earp escaped unharmed. Virgil and Morgan Earp each sustained bullet wounds while Doc Holliday was merely grazed by a bullet to the hip. Ike Clanton, who started the whole incident, was unarmed and allowed to abscond from the scene. Billy Claibourne also fled the scene without participating in the shooting.

Although no one was certain who started the gunfight, in the aftermath of the killings the main question was whether the Earps and Doc Holliday had used the incident as an excuse to settle old scores and had carried out an unlawful execution, or whether they were lawfully trying to disarm their opponents and only fired on them when they felt their lives were threatened. Opinion in Tombstone was divided equally. The events were described the very next day in the *Tombstone Epitaph*, which came out on the side of the lawmen (Figure 14.11).

Following the shootout, the Earps and Doc Holliday went home to tend their wounds. The funerals of Frank and Tom McLaury and of Billy Clanton took place the next day and were attended by a large gathering of angry relatives and friends, vociferous that the deaths were unlawful. On October 28 an inquest was held by the county coroner, where both sides put forward their arguments. The coroner's report issued on October 29 was inconclusive and only stated that the three died from the effects of wounds inflicted by Virgil, Morgan and Wyatt Earp and Doc Holliday, but without apportioning blame.

DAILY EPITAPH

Thursday Morning .... Oct 27, 1881

LOCAL SPLINTERS.

KNIGHTS OF PYTHIAS—meet to night at the court room at 7 30, for regular drill.

THE City Council will meet to day as a Board of Equalization from 10 a m to 2 p. m. All persons wishing to correct their assessment for city taxes will please call on the Board. The Board will sit daily until Nov. 1st.

WE call attention to Charles Glover & Co's advertisement, in another column, which is only surpassed by their sign upon the reservoir on Comstock Hill All that is promised in this their latest proclamation, will be fully sustained upon investigation at their store.

THE Tombstone W. M. & L. Co's reservoir on the summit of Comstock Hill is a conspicuous landmark for miles around. It can be seen from the mesa back of Contention. What makes it more conspicuous is the big sign of Charles Glover & Co, in letters of white upon the red back ground of the tank, of a size that enables one to read it from Third street. This is what we call mammoth advertising. Glover & Co. do not propose to be outdone, either in the

YESTERDAY'S TRAGEDY.

Three Men Hurled Into Eternity in the Duration of a Moment.

The Causes that Led to the Sad Affair.

Stormy as were the early days of Tombstone, nothing ever occurred equal to the event of yesterday. Since the retirement of Ben Sippy as marshal and the appointment of V. W. Earp to fill the vacancy, the town has been noted for its quietness and good order. The fractious and formerly much dreaded cow boys when they came to town were upon their good behavior, and no unseemly brawls were indulged in, and it was hoped by our citizens that no more such deeds would occur as led to the killing of Marshal White, one year ago It seems that this quiet state of affairs was but the calm that precedes the storm that burst in all its fury yesterday, with this difference in results, that the lightning's bolt struck in a different quarter than the one that fell one year ago

deceived me; you told me those men were disarmed; I went to disarm them."

This ends Mr. Coleman's story, which in the most essential particulars has been confirmed by others. Marshal Earp says that he and his party met the Clantons and McLowrys in the alley way by the McDonald place; he called to them to throw up their hands, that he had come to disarm them Instantaneously Bill Clanton and one of the McLowrys fired, and then it became general Mr. Earp says that it was the first shot from Frank McLowry that hit him In other particulars his statement does not materially differ from the statement above given. Ike Clanton was not armed and ran across to Allen street and took refuge in the dance house there The two McLowrys and Bill Clanton all died within a few minutes after being shot The Marshal was shot through the calf of the right leg, the ball going clear through. His brother Morgan was shot through the shoulders, the ball entering the point of the right shoulder blade, following across the back, shattering off a piece of one of the vertebrae and passing out the left shoulder in about the same position that it entered the right. This wound is dangerous but not necessarily fatal, and Virgil's is far more painful than dangerous Doc Holliday was hit upon the scabbard of his pistol, the leather

Figure 14.11 Report in *Tombstone Epitaph* immediately after the shooting, 27 October 1881. *Source*: Courtesy US Library of Congress.

Following a complaint from Ike Clanton, warrants were issued on October 29 for the arrest of the Earps and Doc Holliday to face charges that they had deliberately killed the McLaury brothers and Billy Clanton. While the wounds sustained by Virgil and Morgan Earp were sufficiently serious to preclude them attending the trial, Wyatt Earp and Doc Holliday were arrested and ordered to attend a hearing on October 31.

The charge laid against Wyatt Earp and Doc Holliday was one of unlawful killing and that, when they encountered the McLaurys and the Clantons and asked them to raise their hands, the latter were unarmed and had no intention of initiating a gunfight. Although Tom McLaury moved his hands towards his coat lapel, it was to show that he was unarmed and was not about to draw his gun. It was claimed that at this moment the Earps and Doc Holliday opened fire first and killed them in cold blood as an act of vengeance for troubles past and for threats against their lives. Sheriff Behan supported the charge, but this might have been because he was trying to deflect the accusations against himself of poor leadership in the events leading up to the gunfight and because of previous difficulties between himself and Wyatt Earp.

The subsequent trial was widely reported in the newspapers throughout America, establishing the reputation of Doc Holliday and the Earps against the background of lawlessness in the West. It was seen in terms of good fighting evil, although the 'good' in this case were not all shining white angels.

The Earps and Doc Holliday appointed one of the best lawyers in the county, Thomas Fitch, and though the early stages went against them, he deployed an unusual strategy that allowed Wyatt Earp on November 16 to read from a prepared statement without being cross-examined. Wyatt gave a detailed account of the events leading up to the gunfight, indicating that he and his brothers represented the law and that they only fired their guns after Billy Clanton and Frank McLaury drew theirs first. The defence did not put Doc Holliday on the witness stand, thinking that digging up his colourful past might have been detrimental to their overall case.

After a trial in which the evidence from many witnesses from both sides was heard, the judge issued his verdict on November 30. He found that, with regard to the provocation they were under and the general lawlessness that they were fighting, the Earps and Doc Holliday were not guilty of unlawful killing. In hindsight, the Earps should not have had Doc Holliday with them, bearing in mind the bad feeling that existed between him and Ike Clanton. Indeed, if he had not been present at the time, perhaps the gunfight at the O.K. Corral might never have taken place.

Despite their acquittal, the surviving family and friends of the McLaurys and Clantons swore to get revenge. This they achieved within a short space of time. First, Vigil Earp was ambushed and seriously wounded, his left arm completely shattered. Soon after, Morgan Earp was assassinated.

Wyatt Earp and Doc Holliday, joined by a group of supporters, set out on a murderous vendetta aimed at revenge on those they thought responsible for shooting Virgil and Morgan Earp. During a 3-week period, they murdered at least three people. With warrants out for their arrest, they knew they could never return to Arizona and fled to neighbouring states.

Doc Holliday went to Colorado and remained there until his death. Early on he was arrested while the authorities in Tombstone tried to get him returned and indicted for murder, but this was refused and he was freed. From 1882 until he died, Doc Holliday resided mainly in the towns of Denver and Leadville. In poor health and slowly wasting away from tuberculosis, he continued gambling and drinking, only just staying on the right side of the law. On 8 November 1887 at the age of 36 years, nursed at the end by Big Nose Kate, Doc Holliday died in Glenwood Springs, Colorado.

Although the gunfight at the O.K Corral was soon forgotten, it was well into the twentieth century with the appearance of books and films of the incident that Doc Holliday's legendary status was established. At a time when lawlessness went hand-in-hand with opening up the Wild West, his lack of fear in joining up with the Earps put him at the centre of the most publicised gunfight of that era. Had he stuck with dentistry, one thing is certain: by now he would be totally forgotten.

Doc Holliday has a connection with another person mentioned in this book. In discussing the anaesthetic properties of ether, mention was made of Dr Crawford Long. who, although recorded as having been the first to use it in a small number of pain-free, surgical operations, only published his findings belatedly in 1849 after learning of Morton's discovery (Chapter 3, page 54). Dr Long was Doc Holliday's cousin.

15 Two Famous People with Dental Connections: Paul Revere (American Patriot) and Bernard Cyril Freyberg (Soldier)

Two of the most famous people with strong dental connections are Paul Revere and Bernard Freyberg. Paul Revere's exploits are known to every American, yet he is little known to the rest of the world. He must be the only 'dentist' whose image has appeared on a postage stamp, not once but twice: one a full portrait, the other commemorating the poet Henry Wadsworth Longfellow, in which Revere is seen in the background on a horse (Figure 15.1).

Bernard Freyberg is now virtually unknown outside of his homeland of New Zealand, yet if his life story were written as a novel, his adventures would be considered unbelievable.

Paul Revere (1734–1818)

Paul Revere (Figure 15.2) was born in Boston at the end of 1734. His father (original name Rivoires) was a French Huguenot who, like many others, had fled France to avoid religious persecution by the Catholic state. The town of Boston at that time had a population of about 15,000, made up of many people, such as Calvinists, Puritans and Quakers, who were also fleeing persecution. The people were proud of their independence and desired self-government, inevitably leading to conflict with Britain, whose Parliament and King George III, governed them from London.

Revere's father was a silversmith, and on his death in 1754, Paul, then aged 19, followed in his father's profession and took over the family business. During this early period in the colonisation of North America, there was conflict between England and France, with both countries vying for dominance, as well as fighting with the indigenous Indian population. In 1756, Revere enlisted in the British Army and saw action against the French.

After his return to Boston, Revere completed his apprenticeship, became a master silversmith and married in 1757. His work was much in demand for the quality of his craftsmanship. During his lifetime he was to create more than 5000 silver

Nothing but the Tooth. DOI: http://dx.doi.org/10.1016/B978-0-12-397190-6.00015-8

Figure 15.1 (A) Paul Revere 1958 commemorative U.S. stamp. (B) Henry Wadsworth Longfellow 2007 commemorative U.S. stamp, with Paul Revere on horseback in the background.
Source: (A) from http://commons.wikimedia.org/wiki/File:Paul_Revere_1958_Issue-25c.jpg. This work is in the public domain in the United States because it is a work of the United States Federal Government under the terms of Title 17, Chapter 1, Section 105 of the U.S. Code.

Figure 15.2 Paul Revere.
Source: Photograph ©2012 Museum of Fine Arts, Boston.

items, producing some of the most valuable pieces of the period. One of his finest was the 'Liberty Bowl', reproductions of which are commonly called 'Revere' bowls. It was commissioned in 1768 and commemorates the refusal of the Massachusetts legislature to support a British tax. This was created in response to an early episode in the fight for independence.

In 1761, in order to increase and diversify his income, Revere's skills were turned towards copperplate engraving, which he soon mastered. These engravings became very popular and were produced in very large numbers throughout his life, many with a political slant created as propaganda against British rule. Although somewhat unusual for an artisan, through his many diverse talents, clients and freemasonry, Revere became a well-known and respected figure in Boston society, mixing with influential political figures such as Dr Joseph Warren, John Hancock and Sam Adams, all of whom were to figure prominently in the forthcoming struggle for American independence. Revere was an early activist in this movement, believing strongly that people had the right to be governed by laws of their own making.

Even though he was a master silversmith and engraver, the economy was struggling and Revere needed an additional source of income in 1768 to support his growing family. His solution was to enter the dental profession. This came about through his friendship with an English dentist named John Baker. Little is known about Baker's early life or training, although he seems to have arrived in America in the early 1760s and was one of the earliest practitioners to call himself a 'dental surgeon'. At that time, dentistry was a very primitive occupation, and the procedures on offer would have been tooth extraction, cleaning teeth and fitting false teeth. Baker is known through an advertisement placed in the *Boston Gazette* of 22 January 1768, informing patients that he would be leaving. It seems that Baker taught Revere how to replace natural teeth lost by trauma or disease with false ones. This entailed carving false teeth from ivory (mainly from hippopotamus ivory; see Chapter 2) and attaching them to existing teeth by wire or silk.

When Baker left Boston, Revere advertised his services in the *Boston Gazette* in September 1768 (Figure 15.3). It reads:

> *Whereas many persons are so unfortunate as to lose their foreteeth by accident, or other ways, to their great detriment, not only in looks, but speaking, both in public and private: this is to inform all such, that they may have them replaced with false ones, that look as well as the natural, and answer the end of speaking, to all intents, by Paul Revere, Goldsmith, near the Head of Dr Clarke's wharf, Boston. All persons who have had false teeth fix'd by John Baker, Surgeon Dentist, and they have got loose (as they will in time) may have them fastened by the abovesaid Revere, who learnt the method of fixing them from Mr Baker.*

Nearly 2 years later, he placed another announcement in the *Boston Gazette* in July 1770 (Figure 15.4). Under the heading 'Artificial Teeth' it reads:

> *Paul Revere takes this method of returning his most sincere thanks to the gentlemen and ladies who have employed him in the care if their teeth, he would now inform them and all others, who are so unfortunate as to lose their teeth by accident or otherwise that he still continues the business of a dentist, and flatters himself that from the experience he has had these two years, (in which time he has fixt some hundreds of teeth) that he can fix them as well as any surgeon dentist who ever came from London, he fixes them in such a manner that they are not only an*

ornament but of real use in speaking and eating. He cleanses the teeth and will wait on any gentleman or lady at their lodgings, he may be spoke with at his shop opposite Dr Clark's at the North End, where the gold and silversmith's business is carried on in all its branches

Because Revere's ledgers show items of dentistry up to 1774, it is clear that the practise of dentistry was a major occupation for him for at least 6 years. His patients even included his friend, Dr Joseph Warren, for whom he must have fixed a false tooth in the upper left canine region by gold wire (see page 215).

While he was practising dentistry, Revere may have indirectly benefited the profession by influencing the children of two neighbours, who were later to have a big

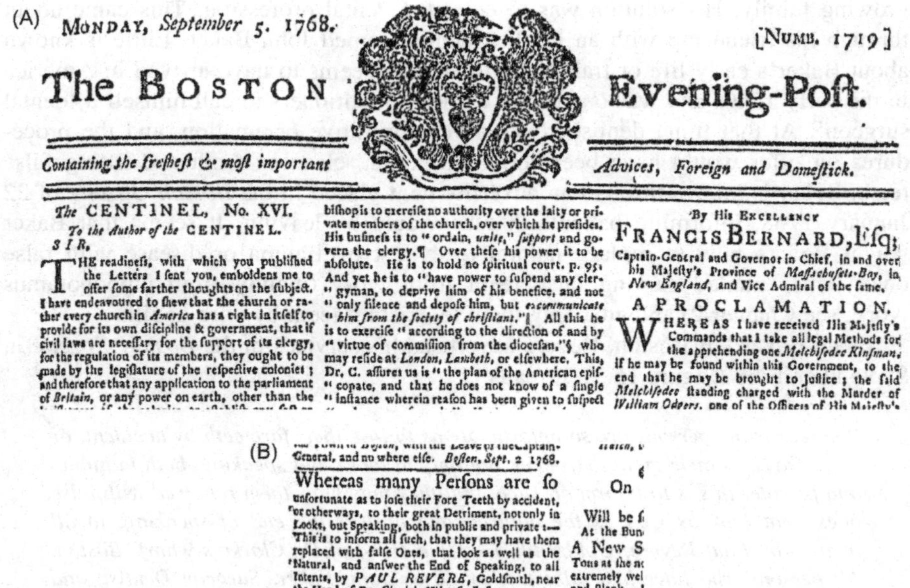

(A) MONDAY, *September* 5. 1768. [NUMB. 1719]

The BOSTON Evening-Post.

Containing the freshest & most important Advices, Foreign and Domestick.

The CENTINEL, No. XVI.
To the Author of the CENTINEL.
SIR,

THE readiness, with which you published the Letters I sent you, emboldens me to offer some farther thoughts on the subject. I have endeavoured to shew that the church or rather every church in *America* has a right in itself to provide for its own discipline & government, that if civil laws are necessary for the support of its clergy, for the regulation of its members, they ought to be made by the legislature of the respective colonies; and therefore that any application to the parliament of *Britain*, or any power on earth, other than the ...

bishop is to exercise no authority over the laity or private members of the church, over which he presides. His business is to "ordain, *unite*," *support* and govern the clergy.¶ Over these his power it to be absolute. He is to hold no spiritual court. p. 95. And yet he is to "have power to suspend any clergyman, to deprive him of his benefice, and not only silence and depose him, but *excommunicate him from the society of christians*."‖ All this he is to exercise "according to the direction of and by virtue of commission from the diocesan,"§ who may reside at *London*, *Lambeth*, or elsewhere. This, Dr. C. assures us is "the plan of the American episcopate, and that he does not know of a single instance wherein reason has been given to suspect ...

By His EXCELLENCY
FRANCIS BERNARD, Esq;
Captain-General and Governor in Chief, in and over his Majesty's Province of *Massachusetts-Bay*, in *New England*, and Vice Admiral of the same.
A PROCLAMATION.
WHEREAS I have received His Majesty's Commands that I take all legal Methods for the apprehending one *Melchisedec Kinsman*, if he may be found within this Government, to the end that he may be brought to Justice; the said *Melchisedec* standing charged with the Murder of *William Odgers*, one of the Officers of His Majesty's ...

(B) ... General, and no where else. *Boston, Sept.* 2 1768.

Whereas many Persons are so unfortunate as to lose their Fore Teeth by accident, or otherways, to their great Detriment, not only in Looks, but Speaking, both in public and private:— This is to inform all such, that they may have them replaced with false Ones, that look as well as the Natural, and answer the End of Speaking, to all Intents, by PAUL REVERE, Goldsmith, near the Head of Dr. *Clarke's* Wharf, *Boston*.——

*** All Persons who have had false Teeth fix'd by Mr. *John Baker*, Surgeon Dentist, and they have got loose (as they will in Time) may have them fastened by the abovesaid *Revere*, who learnt the Method of fixing them from Mr. *Baker*.

Whereas the Assessors are required by Law to give seasonable Warning to the Inhabitants to give and bring in to the Assessors, true and perfect Lists of their Polls and rateable Estates, they are possessed of: In Conformity to the Law, The Assessors of the Town of *Boston* do hereby notify the Inhabitants of said Town, that they shall attend at their Chamber at the East End of Faneuil Hall Market, to receive the same from Monday the 5th to the 19th Instant, at the usual Hours, on every Day in the Week, except Lord's-Day and Saturday
By Direction of the Assessors.
Boston, September 1. 1768.

Figure 15.3 (A) Header of Boston Gazette for September 1768. (B) Advert placed by Paul Revere.
Source: Courtesy of U.S. Library of Congress.

impact on the American dental profession. One, Isaac Greenwood, had four sons, all of whom became dentists. John Greenwood supplied George Washington's dentures and invented the first dental drill in 1790. The other neighbour was Josiah Flagg. His son, Josiah Flagg Jr (born 1763), was one of the first American-born dentists and may have constructed the earliest authentic dental chair in North America. He was also the first in a dynasty of dentists.

Politically, the 13 colonies in America had virtually no representation in the British Parliament, which made the laws and enforced the taxes. From the late 1760s, Boston was the centre for the colonists' struggle for more independence from Britain. In 1768, Britain felt it necessary to send over extra troops to Boston for the purposes of intimidation and enforcement of legislation, including the payment of unpopular taxes. This move was seen by the colonists as inflammatory and stirred up unrest even further. The opposition was led by acquaintances of Revere such as James Otis, John Hancock and Dr George Warren, who realised the importance of forming links with other states and keeping them abreast of developments. Revere was completely sympathetic to, and an active supporter of, their views. Although never one for writing and making speeches, he was identified as a man of ability and integrity and, most importantly, an organiser who could get things done. Especially relevant was his unrivalled list of contacts in Boston, built up across all classes, in his capacity as a respected artisan, freemason and churchgoer.

In 1770, a small gathering of British troops, subjected to some hostility from the locals, fired into the crowd, killing five civilians. This incident, known as the

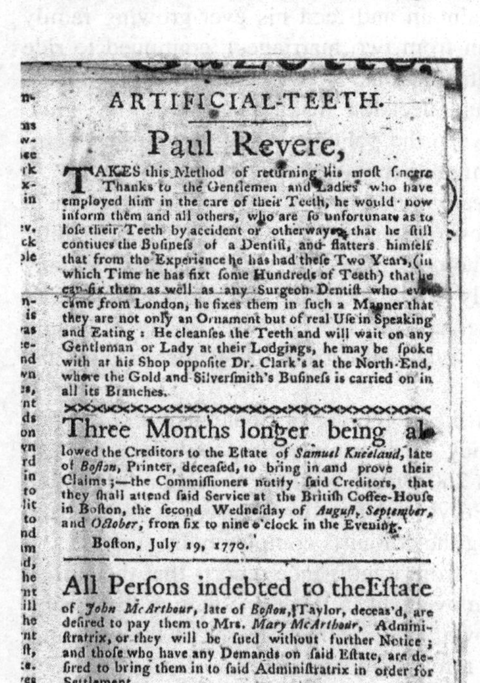

ARTIFICIAL-TEETH.

Paul Revere,

TAKES this Method of returning his most sincere Thanks to the Gentlemen and Ladies who have employed him in the care of their Teeth, he would now inform them and all others, who are so unfortunate as to lose their Teeth by accident or otherways, that he still contiues the Business of a Dentist, and flatters himself that from the Experience he has had these Two Years, (in which Time he has fixt some Hundreds of Teeth) that he can fix them as well as any Surgeon-Dentist who ever came from London, he fixes them in such a Manner that they are not only an Ornament but of real Use in Speaking and Eating: He cleanses the Teeth and will wait on any Gentleman or Lady at their Lodgings, he may be spoke with at his Shop opposite Dr. Clark's at the North-End, where the Gold and Silversmith's Business is carried on in all its Branches.

Three Months longer being allowed the Creditors to the Estate of *Samuel Kneeland*, late of *Boston*, Printer, deceased, to bring in and prove their Claims;—the Commissioners notify said Creditors, that they shall attend said Service at the British Coffee-House in Boston, the second Wednesday of *August*, *September*, and *October*, from six to nine o'clock in the Evening.

Boston, July 19, 1770.

All Persons indebted to the Estate of *John McArthour*, late of *Boston*, Taylor, deceas'd, are desired to pay them to Mrs. *Mary McArthour*, Administratrix, or they will be sued without further Notice; and those who have any Demands on said Estate, are desired to bring them in to said Administratrix in order for Settlement.

Figure 15.4 Advert placed by Paul Revere in *Boston Gazette*, July 1770.
Source: Courtesy of U.S. Library of Congress.

Boston Massacre, led Revere to produce an engraving of a powerful illustration by Paul Walker, widely used as propaganda against the British.

The next major incident occurred in December 1773 when government officials tried to enforce a punishing tax on tea (to maintain the monopoly of this trade held by the East India Company). When three ships carrying a cargo of tea arrived in Boston Harbour, the colonists refused to pay the tax and demanded the tea be returned directly to Britain. Their slogan was 'No taxation without representation'. Thousands of agitated townspeople gathered around Boston Harbour, and in the evening of December 16, a group of between 100 and 150 men boarded the ships, partly disguised as Mohawk Indians, and emptied all 342 tea chests into the water. Among the men was Paul Revere.

Following this revolt, and with signs of conflict growing daily, the colonist leaders in Boston felt it essential for the other states to be kept informed of what was happening in all parts of the country. Paul Revere was chosen for the important task of delivering the news of the revolt to the neighbouring cities of Philadelphia and New York. This he did in the only way then possible, on horseback. He covered the 700-mile round-trip in just 11 days. In the following 2 years, he would make at least five other similar journeys.

It was obvious that an intelligence network was required to spy and report on the activity of British troops. In addition, militia had to be organised and trained to be ready at very short notice (hence the term 'minutemen') to defend their homes and livelihoods against further British incursions. Revere, whilst still pursuing his varied business interests necessary to maintain and feed his ever-growing family (in his long life, he produced 16 children from two marriages), continued to ride back and forth with important news and instructions. He was more than just a messenger. He was involved in planning and decision making and was the 'glue' between disparate groups, especially between the politicians and the artisans, helping to produce a more cohesive force. While others talked, he acted!

Although their military forces were greatly outnumbered, the British government did not believe that the colonists could form a unified force to defeat their regular troops under the command of General Thomas Gage. The view was that as many of the colonists were British, they surely would not take up arms against their homeland. However, Britain would underestimate the colonists, many of whom already had experience in battle, having fought alongside the British against the French and local Indians.

The next major incident took place at the beginning of September 1774, when General Gage started to enforce the policy of disarming the population. He sent Lieutenant-Colonel George Madison with 260 British regulars to seize a large store of gunpowder from the Massachusetts Provincial Powder House, six miles from Boston. This proved a total success, taking the colonists completely by surprise.

By April 1775, it was abundantly clear from intelligence reports that the British were bent on crushing the independence movement in Boston, although the precise date of this intended action was not known. The colonists learnt of this not only from sympathisers in Britain but, more remarkably, probably also from General Gage's own wife Margaret, who was American by birth. The plan was to arrest two of the leaders, Dr Sam Adams and John Hancock, at Lexington some 12 miles away. Following this, British forces were to march on a further six miles to the town of Concord and destroy the colonists' main arms depot. General Gage did not think the colonists would resist. On learning of the plan, Dr Warren sent Revere to Concord on April 15 to warn them of a possible raid and instruct them to hide their arms and gunpowder.

There were two alternative routes the British could take: a longer, overland one or a shorter, sea route after being ferried across the nearby Charles River. Once the route was known and the British troops prepared to march, it was to be signalled using lanterns from the highest vantage point, the tower of Christ Church, to colonist lookouts across the river in Charlestown: one light for the land route and two lights for the sea route.

All the intelligence eventually pointed to a river crossing, commencing on the evening of April 18. Two lights were signalled for a brief moment from the tower of Christ Church. With the knowledge that about 700 British troops were beginning to move on Lexington by the sea route, Dr Warren instructed Revere to ride to Lexington. He was to take the quicker, sea route, and in case he did not succeed, another rider, William Davies, was sent by the longer, land route. The mission of both riders was to spread the alarm that the British regulars were on the way and to help Sam Adams and John Hancock avoid arrest. No instructions were given to them to ride on to Concord. At 10 pm on the evening of July 18, Revere left to carry out his regular duty of messenger, unaware that his journey would establish his place in American history. Although many may think that his epic ride to warn of the British advance was carried out by him single-handedly, at every stage along the route others were incorporated in spreading the alarm.

Revere was first rowed across the narrow Charles River on a clear moonlit night, under the 64 guns of the British warship Somerset. Luckily, he was not spotted and landed at Charlestown at 11 pm. Here, other friends provided him with a fine, strong horse named Brown Beauty. Encountering a British patrol deliberately deployed to prevent messages of British intent getting through, Revere was able to outride them and escape. He arrived at Lexington at 12 pm, warning people along the route, who spread the message by additional riders, church bells, gunfire and bonfires that the British regulars were coming. The colonist militias were rapidly mustered and rushed to defend Lexington and Concord and, if necessary, to fight the British.

Having arrived at Lexington, Revere passed on the warning to Samuel Adams and John Hancock that they should flee from the British. Further alarms were spread widely from Lexington.

William Dawes, the second rider, arrived in Lexington soon after Revere. History shows that, compared to Revere, his ride had little effect in spreading the alarm. Dawes was not a 'connector', lacking Revere's rare social gifts and list of contacts, a phenomenon identified in Malcolm Gladwell's book *The Tipping Point*. Revere's unique role was the result of his long association with the politicians of Massachusetts: he knew who they were and where to find them, and they in turn had complete trust in his word. One can only surmise how history might have been affected had Revere not reached Lexington.

Just before 1 am, Revere and Dawes decided to ride on to Concord, to reinforce the warning of the oncoming British troops. They were joined by a third rider, Dr Samuel Prescott, but shortly after leaving Lexington, they were apprehended by a British patrol. Dawes and Prescott evaded capture and Prescott succeeded in reaching Concord, where further riders were despatched. Revere was eventually released about 3 am, but his horse was confiscated. He walked back to Lexington without ever reaching Concord. Brown Beauty was never seen again.

On returning to Lexington, Revere found Adams and Hancock still there and had to persuade them that they were too important to take part in the fight. He escorted them on their way out of Lexington to seek shelter in the nearby countryside. Revere's final task was to rescue Hancock's heavy trunk, which was full of important documents. It was about 4:30 am as he carried out this rescue that he witnessed the early stages of the next dramatic event.

The advanced column of British troops, numbering about 400, reached Lexington Village Common, with the officers in the rear. At this time, about 60 to 70 armed militia were present in Lexington under the command of Captain John Parker. They had no plan of action but decided to confront the British rather than retreat. It was important to the colonists in any ensuing struggle that they should not be the ones to fire first. In this way, if fighting broke out, they would not be morally responsible for starting it.

The British troops were headed by an inexperienced marine officer, Lieutenant Jesse Adair. Marching towards the small detachment of colonist militia ahead of him, he had the option of turning left and avoiding them or turning right and meeting them head on. In the absence of anyone senior, he made the fateful decision to turn right in battle formation, seemingly ready to confront the colonists. He was joined by some of his own mounted officers. With the two forces now separated by only about 20 m, a cool head at the front of the British column might still have been able to take control and diffuse the situation before anything serious happened. At this critical moment, a shot rang out. No one knows who fired it, but the response of the young and nervous British regulars was to begin firing, even though no order had been given. There was little return of fire by the colonist militia, who rapidly dispersed from the scene.

When news of the shooting reached farther down the British line to its leader, Colonel Smith, he immediately ordered his troops to cease fire. The result of this one-sided action was that seven militia men were killed and nine wounded: the British suffered just one soldier wounded. Although none of the participants was aware of its significance at the time, in historical terms this incident marks the first battle in the American War of Independence. It was now about 5 am on April 19.

Following the clash on Lexington Village Common, the British troops searched in vain for Adams and Hancock. They then moved on to Concord to seek out arms and ammunition, the main purpose of the mission. Now that the advantage of surprise had been lost, they realised their chances of success were low. With growing evidence of the increasing numbers of opposition forces, Smith sent back to Boston for reinforcements.

Earlier that morning, a messenger had been dispatched from Concord to Lexington to assess the progress of the British troops, and after witnessing the battle, he returned to alert the town. As the British marched towards Concord, initially they outnumbered the militia. The town was evacuated but the militia held a strong defensive position in the surrounding hills. On arriving in Concord, the British regulars had little success in their search for weapons and gunpowder. By now militia numbers had grown considerably, and in one area they outnumbered a British detachment of about 100 troops guarding a bridge. They advanced on this position, and when the British troops fired first, killing one or two of them, they returned this fire, killing and injuring a number of British troops. Most remarkably, the militia witnessed the British troops retreating in disarray.

Colonel Smith, realising the growing gravity of the situation, organised a retreat of his forces back to Lexington. All along the route they were followed by an ever-growing number of colonist militia, whose confidence in their ability to fight the British had now risen. As they knew every feature of the landscape, they were able to ambush the retreating forces at strategic points. They aimed particularly for British officers, readily distinguished by the bright red colour of their uniforms, and these higher ranks suffered grievously during the retreat. On the road to Lexington, Captain Parker and his militia, who were the first to shed blood in the conflict, had the opportunity to redeem themselves by killing a number of British troops and wounding Colonel Smith.

With the outcome looking desperate for the outnumbered British troops as they neared Lexington, they were rescued by the relief party they had requested earlier that morning. It consisted of a large force of about 1000 troops led by Brigadier Lord Hugh Percy. It was now about 2:30 pm.

For the rest of the day, the British slowly retreated back to Charlestown, across the river from Boston, arriving as night fell. The area was surrounded by colonist forces. British casualties that day numbered nearly 300, including over 70 killed. The colonist losses were less than a hundred, with about 50 killed. The British view that the colonists would not fight had proved disastrously wrong, as they did fight and very capably.

Revere's dental knowledge was of use once more during this period. While besieging the British forces in Boston, the colonists received information that the British intended to occupy the high ground around Bunker Hill. In response to this, the colonists occupied it first, but on June 17 the British forces drove the colonists away, although at the expense of considerable casualties. Dr Joseph Warren was killed and buried in a mass grave. The following year, his family and friends, including Revere, visited the burial site in the hope of identifying him so that they could provide a more appropriately marked grave. This seemed an almost impossible task in view of the decomposition of the bodies. However, Revere was able to identify Warren's body by recognising his own previous dental work, when he had attached a false tooth in the upper left canine region by means of gold wire. This is regarded as one of the earliest examples of forensic odontology, where a body was identified by dental evidence.

During the American War of Independence, which lasted until 1783, although not contributing militarily, Revere used his abilities in other ways. His skill as an engraver was employed in printing money, and he also helped establish a factory for making gunpowder. After the war, he became a very successful and wealthy entrepreneur and businessman. In addition to his continued work as a silversmith, he established a major iron and bell foundry and the first factory producing copper sheeting in North America.

Paul Revere died at the age of 83 in 1818. From newspaper obituaries, it can be seen that he was well known and respected around Boston. An obituary printed in the *New England Galaxy and Masonic Magazine* on 15 May 1818 (Figure 15.5A), reads:

> *On Sunday departed this life, PAUL REVERE, Esq. in the 84th year of his age. During his protracted life, his activity in business and benevolence, the vigour of his mind, and strength of his constitution were unabated. He was one of the earliest and most indefatigable Patriots and Soldiers of the Revolution, and has filled with fidelity, ability and usefulness, many important situations in the military and civil service of his country, and at the head of valued and beneficent institutions. Seldom has the tomb closed upon a life so honourable and useful.*

(A)

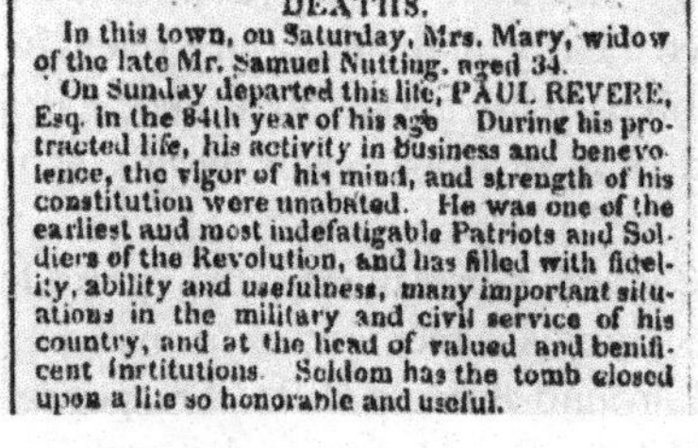

DEATHS.

In this town, on Saturday, Mrs. Mary, widow of the late Mr. Samuel Nutting, aged 34.

On Sunday departed this life, PAUL REVERE, Esq. in the 84th year of his age During his protracted life, his activity in business and benevolence, the vigor of his mind, and strength of his constitution were unabated. He was one of the earliest and most indefatigable Patriots and Soldiers of the Revolution, and has filled with fidelity, ability and usefulness, many important situations in the military and civil service of his country, and at the head of valued and benificent institutions. Seldom has the tomb closed upon a life so honorable and useful.

(B)

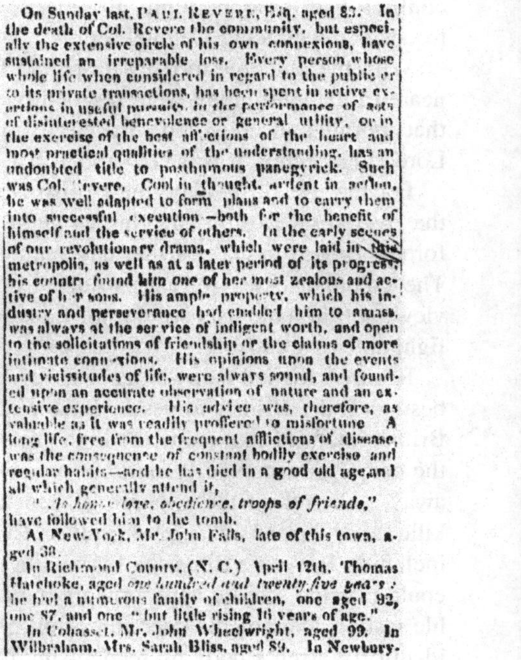

On Sunday last, PAUL REVERE, Esq. aged 83. In the death of Col. Revere the community, but especially the extensive circle of his own connexions, have sustained an irreparable loss. Every person whose whole life when considered in regard to the public or to its private transactions, has been spent in active exertions in useful pursuits, in the performance of acts of disinterested benevolence or general utility, or in the exercise of the best affections of the heart and most practical qualities of the understanding, has an undoubted title to posthumous panegyrick. Such was Col. Revere. Cool in thought, ardent in action, he was well adapted to form plans and to carry them into successful execution—both for the benefit of himself and the service of others. In the early scenes of our revolutionary drama, which were laid in this metropolis, as well as at a later period of its progress, his country found him one of her most zealous and active of her sons. His ample property, which his industry and perseverance had enabled him to amass, was always at the service of indigent worth, and open to the solicitations of friendship or the claims of more intimate connexions. His opinions upon the events and vicissitudes of life, were always sound, and founded upon an accurate observation of nature and an extensive experience. His advice was, therefore, as valuable as it was readily proffered to misfortune. A long life, free from the frequent afflictions of disease, was the consequence of constant bodily exercise and regular habits—and he has died in a good old age, and all which generally attend it,

"As honor, love, obedience, troops of friends,"

have followed him to the tomb.

At New-York, Mr. John Falls, late of this town, aged 38.

In Richmond County, (N. C.) April 12th, Thomas Hatchoke, aged *one hundred and twenty five years*: he had a numerous family of children, one aged 92, one 87, and one "but little rising 16 years of age."

In Cohasset, Mr. John Wheelwright, aged 99. In Wilbraham, Mrs. Sarah Bliss, aged 89. In Newbury-

Figure 15.5 (A) Obituary of Paul Revere as printed in the *New England Galaxy and Masonic Magazine*, 15 May 1818, **1**(31): 3. (B) Obituary of Paul Revere as printed in the *Boston Intelligencer, and Morning and Evening Advertiser*, 16 May 1818, **4**(40):2. *Source*: Courtesy of U.S. Library of Congress.

A second and longer obituary printed in the *Boston Intelligencer, and Morning and Evening Advertiser*, on 16 May 1818 (Figure 15.5B), reads:

> *On Sunday last, Paul Revere, Esq., aged 83. In the death of Col. Revere the community, but especially the extensive circle of his own connexions, have sustained an irreparable loss. Every person whose whole life when considered in regard to the public or to its private transactions, has been spent in active exertions in useful pursuits in the performance of acts of disinterested benevolence or general utility, or in the exercise of the best affections of the heart and most practical qualities of the understanding, has an undoubted title to posthumous panegyrick. Such was Col. Revere. Cool in thought, ardent in action, he was well adapted to form plans and to carry them into successful execution – both for the benefit of himself and the service of others. In the early scenes of our revolutionary drama, which were laid in this metropolis, as well as at a later period of its progress, his country found him one of her most zealous and active of her sons. His ample property, which his industry and perseverance had enabled him to amass, was always at the service of indigent worth, and open to the solicitations of friendship or the claims of more intimate connexions. His opinions upon the events and vicissitudes of life, were always sound, and founded upon an accurate observation of nature and an extensive experience. His advice was, therefore, as valuable as it was readily proffered to misfortune. A long life, free from the frequent afflictions of disease, was the consequence of constant bodily exercise and regular habits – and he has died in a good old age, and all which generally attend it. As honour, love, obedience, troops of friends have followed him to the tomb.*

However, outside of Boston, Revere was little known compared with other military and political figures who successfully fought the British. Without Henry Wadsworth Longfellow, his name surely would have lapsed into obscurity. Longfellow was America's most famous poet following the publication in 1855 of his epic poem *The Song of Hiawatha*. In January 1861, just before the outbreak of the American Civil War, he published his poem *Paul Revere's Ride*. This work described Paul Revere's midnight ride and how it galvanised the people in the surrounding countryside (but romanticising events and distorting history for poetic needs, such as having him ride successfully on to Concord). It gave all the glory to Revere in initiating the uprising, while giving little credit to all the others who spread the alarm. The poem struck a deep chord in the American people during and after the Civil War because it reminded them of the unity of their nation nearly a hundred years earlier when successfully facing down the repressive forces affecting their civil liberties. It elevated Paul Revere to the pantheon of American patriots. Here is the poem:

Paul Revere's Ride
by
Henry Wadsworth Longfellow

Listen my children and you shall hear
Of the midnight ride of Paul Revere,

On the eighteenth of April, in Seventy-five;
Hardly a man is now alive
Who remembers that famous day and year.

He said to his friend, 'If the British march
By land or sea from the town to-night,
Hang a lantern aloft in the belfry arch
Of the North Church tower as a signal light,–
One if by land, and two if by sea;
And I on the opposite shore will be
Ready to ride and spread the alarm
Through every Middlesex village and farm,
For the country folk to be up and to arm'.

Then he said 'Good-night!' and with muffled oar
Silently rowed to the Charlestown shore
Just as the moon rose over the bay,
Where swinging wide at her moorings lay
The Somerset, British man-of-war;
A phantom ship, with each mast and spar
Across the moon like a prison bar,
And a huge black hulk, that was magnified
By its own reflection in the tide.

Meanwhile, his friend through alley and street
Wanders and watches, with eager ears,
Till in the silence around him he hears
The muster of men at the barrack door,
The sound of arms, and the tramp of feet,
And the measured tread of the grenadiers,
Marching down to their boats on the shore.
Then he climbed the tower of the Old North Church
By the wooden stairs, with stealthy tread,
To the belfry chamber overhead,
And startled the pigeons from their perch
On the sombre rafters, that round him made
Masses and moving shapes of shade,
By the trembling ladder, steep and tall,
To the highest window in the wall,
Where he paused to listen and look down
A moment on the roofs of the town
And the moonlight flowing over all.

Beneath, in the churchyard, lay the dead,
In their night encampment on the hill,
Wrapped in silence so deep and still
That he could hear, like a sentinel's tread
The watchful night-wind, as it went
Creeping along from tent to tent,
And seeming to whisper, 'All is well!'
A moment only he feels the spell
Of the place and the hour, and the secret dread

Of the lonely belfry and the dead;
For suddenly all his thoughts are bent
On a shadowy something far away,
Where the river widens to meet the bay,
A line of black that bends and floats
On the rising tide like a bridge of boats.

Meanwhile, impatient to mount and ride
Booted and spurred, with a heavy stride
On the opposite shore walked Paul Revere.
Now he patted his horse's side,
Now he gazed at the landscape far and near,
Then, impetuous, stamped the earth,
And turned and tightened his saddle girth;
But mostly he watched with eager search
The belfry tower of the Old North Church,
As it rose above the graves on the hill,
Lonely and spectral and sombre and still.
And lo! as he looks, on the belfry's height
A glimmer, and then a gleam of light!
He springs to the saddle, the bridle he turns,
But lingers and gazes, till full on his sight
A second lamp in the belfry burns.

A hurry of hoofs in a village street,
A shape in the moonlight, a bulk in the dark,
And beneath, from the pebbles, in passing, a spark
Struck out by a steed flying fearless and fleet;
That was all! And yet, through the gloom and the light
The fate of a nation was riding that night;
And the spark struck out by that steed, in his flight,
Kindled the land into flame with its heat.
He has left the village and mounted the steep.
And beneath him, tranquil and broad and deep,
Is the Mystic, meeting the ocean tides;
And under the alders that skirt its edge,
Now soft on the sand, now loud on the ledge,
Is heard the tramp of his steed as he rides.

It was twelve by the village clock
When he crossed the bridge into Medford town.
He heard the crowing of the cock
And the barking of the farmer's dog,
And felt the damp of the river fog,
That rises after the sun goes down.

It was one by the village clock
When he galloped into Lexington.
He saw the gilded weathercock
Swim in the moonlight as he passed.
And the meeting-house windows, black and bare,
Gaze at him with a spectral glare,

As if they already stood aghast
At the bloody work they would look upon.

It was two by the village clock
When he came to the bridge in Concord town.
He heard the bleating of the flock
And the twitter of birds among the trees,
And felt the breath of the morning breeze
Blowing over the meadow brown.
And one was safe and asleep in his bed
Who at the bridge would be first to fall,
Who that day would be lying dead,
Pierced by a British musket ball.

You know the rest. In the books you have read
How the British Regulars fired and fled,
How the farmers gave them ball for ball,
From behind each fence and farmyard wall,
Chasing the redcoats down the lane,
Then crossing the fields to emerge again.
Under the trees at the turn of the road,
And only pausing to fire and load.

So through the night rode Paul Revere;
And so through the night went his cry of alarm
To every Middlesex village and farm.
A cry of defiance, and not of fear,
A voice in the darkness, a knock at the door,
And a word that shall echo for evermore!
For, borne on the night-wind of the Past,
Through all our history, to the last,
In the hour of darkness and peril and need,
The people will waken and listen to hear
The hurrying hoof-beats of that steed,
And the midnight message of Paul Revere.

Note: Henry Wadsworth Longfellow's poem *Paul Revere's Midnight Ride* can be found at the following website: http://poetry.eserver.org/paul-revere.html.

Longfellow can be linked directly to the events described in Chapter 3, namely the discovery of general anaesthesia. On 7 April 1847, his wife, Fanny, became the first woman in the United States to have a child delivered under ether anaesthesia, thereby abolishing much of the pain associated with childbirth. The anaesthetic was administered by the Dean of Dentistry at Harvard University, Nathan Keep, who had gained his experience by using ether to extract teeth.

Bernard Cyril Freyberg (1889–1963)

Acts of great bravery are often acknowledged by the award of a medal. The highest award for bravery in battle in the United Kingdom and the Commonwealth is the

Victoria Cross. It originated during the Crimean War (1853–1856) and, unlike previous awards, was open to all ranks. Queen Victoria herself chose the motto 'For Valour'. Although it is widely believed that the metal for the medal was derived from guns captured during the siege of Sebastopol in the Crimean War, it is now generally accepted that it had a less romantic origin from spare guns found in the Woolwich Arsenal in London.

The level of bravery necessary to receive the Victoria Cross is such that many have been awarded posthumously. The importance and rarity of the medal is evidenced by the recent sale of a Victoria Cross for £400,000, awarded to an Australian soldier during the Gallipoli campaign of 1915. Three people have each received two Victoria Crosses. Two were surgeons in the Royal Army Medical Corps, both for rescuing wounded soldiers while under heavy fire.

One Victoria Cross holder was a dentist, although the award had nothing to do with his profession. This award alone would have climaxed the achievements of the person concerned. However, far from being the high point in his career, it was just the start of an extraordinary journey. He was actively involved in so many murderous battles that it is hard to believe he could have survived First World War let alone Second World War and become one of the most decorated military figures in history. I refer to Bernard Cyril Freyberg, dental surgeon (Figure 15.6).

Freyberg's family came from Germany and had a distinguished record of military service. He was born in England in 1889, the youngest of five children, but due to financial difficulties, his father emigrated to Wellington, New Zealand, in 1891. At school, Freyberg did not shine academically but was very athletic and enjoyed outdoor pursuits. He excelled at swimming and represented New Zealand in this sport. Although nicknamed Tiny as a child, being the youngest and initially the smallest, he grew to be an imposing, powerfully built figure over 6 ft tall. However, the nickname Tiny stuck with him throughout his life.

Due to difficult financial circumstances, Freyberg left school just short of his sixteenth birthday without any qualifications. As this precluded him becoming a doctor, the next best (and cheapest) thing that he could pursue was to become a dentist. At

Figure 15.6 Illustration of Sir Bernard Freyberg, 1952. *Source*: Courtesy of National Army Museum, New Zealand.

that time dentistry was regarded more as a trade than a profession. Prospective candidates would become apprenticed to a dentist for a fee and spend some years in training, finally being examined to assess their proficiency before being awarded a certificate to practise. However, at the very time that Freyberg commenced his apprenticeship with a local dentist, in 1904 a new act was passed that raised the professional standards of dentistry and required a much more stringent and prolonged course of training. This requirement was associated with the establishment of the first Chair of Dentistry in New Zealand at the University of Otago, Dunedin.

Being one of the last candidates to take the old form of registration, Freyberg was suddenly informed that if he did not obtain the registration examination before January 1908, he would have to take the new, university-based examination. With very little time left before this date expired, it was arranged that the few apprentices who were affected could attend a special course run by the new dental school in Dunedin. Although he attended the course, there was insufficient time for Freyberg to be brought up to the required standard. He returned home and determinedly sought his rights by petitioning Parliament to make an exception to the new rules for dental registration. This he achieved, and in May 1911 his name was added to the dental register.

For the next couple of years, Freyberg practised as a dentist in various locations around New Zealand. At the same time, he joined the Territorial Army. It seems that life as a dentist did not fully satisfy him, so he left New Zealand to seek adventure in the world outside. The storm clouds of war that were gathering in Europe seemed of little consequence, and in March 1914 he sailed to America.

One of his first moves on arriving in San Francisco in April 1914 was to travel to Mexico. There was much political unrest in that country, and civil war had broken out. After an adventurous few months there, hostilities in Europe erupted with the outbreak of First World War at the beginning of August 1914. Freyberg had only one thought in mind. He hurriedly left Mexico, travelled to New York, boarded the first ship for Liverpool and made his way to London to volunteer for war duty.

On arriving in London, Freyberg applied to join the Royal Naval Division (RND) and did not hesitate to accost the First Lord of the Admiralty himself, Winston Churchill, in the street, to gain his support. This relationship had important implications later. Due to his prior experience in the Territorial Army in New Zealand, Freyberg was immediately commissioned into the RND. Officers and men in the RND were eventually to include some of the best minds and most influential people of that generation, many lost to the nation, including the poet Rupert Brooke. Meeting and mixing with these people would profoundly influence Freyberg's life. He was soon placed in command of a company of over 200 men. It was September 1914, and he was 25 years old.

At the onset of war, things went badly for Britain and its allies. With all their professional forces already overseas as the British Expeditionary Force (BEF), the British government had no fully trained forces in reserve. To resist the German advance, the only immediately available force was the RND, although they were only partly trained and poorly equipped. Despite this, they were sent to help support the retreating BEF. Arriving to help in the defence of Antwerp, Freyberg saw his first action, which resulted in considerable loss of life and retreat back to England.

In March 1915, Freyberg left England for the ill-fated Gallipoli campaign. It was on a stopover on the Greek island of Skyros that Rupert Brooke, already weakened by a bout of dysentery, suffered an insect bite that became infected, and he died in April 1915. Freyberg, being his company commander, was one of those who buried him on that island.

As the opposing lines of trenches in Europe had stabilised the fighting front with little progress on either side, the general aim of the Gallipoli campaign was to open up a second front and help relieve pressure on Britain's Russian ally. Churchill devised a plan to use a mainly naval force to attack Turkey in the region of the Dardanelle Straits that led into the Sea of Marmara. When the initial attack failed, the plan was hastily revised to include the landing of an army on the beaches of the Gallipoli peninsula. However, from the outset the Gallipoli campaign was doomed to failure as a result of a lack of leadership, poor planning and poor execution. The quality and quantity of the Turkish resistance had been underestimated, and the enemy was given prior warning as to where the invasion would take place. The campaign exposed the combined Australian and New Zealand Army Corps (ANZAC) to their first major conflict in support of the British Empire, and their courageous joint action helped establish their feelings of independence.

On arriving at the Dardanelles, one of the first tasks of the RND was to act as a decoy and try to mislead the Turks as to the real site of the main invasion around Gallipoli. The aim was to undertake diversionary activity farther inland, at a site called Bulair. The initial plan was to land some men at night on the coast in order to light flares and give the impression of a much larger force. Thinking this a somewhat dangerous and unnecessary task for a group of men, Freyberg volunteered to swim on his own to the shore, towing the flares behind him. This he carried out successfully. For this act of bravery, which needed great strength and endurance to swim the distance of approximately two miles in icy water, he was awarded the Distinguished Service Order (DSO).

With the Turkish army and artillery occupying the high ground, the initial landings on the beaches around the Gallipoli peninsula were successfully resisted. There followed months of bloody and brutal warfare, fought with enormous courage on both sides under the harshest of conditions: the blazing heat of summer was followed by the extreme cold of winter. The general conditions the troops found themselves in were appalling. Casualties were very high on both sides, with thousands of lives lost for temporary gains of just a few hundred yards, which were reversed within a short period of time. About 50,000 men died on each side, with double that number wounded.

Although the role of the naval forces was initially anticipated to be a supportive one to the army, the quick reverses suffered meant that Freyberg and the RND soon found themselves in the thick of the land campaign. Within a few short weeks, the ranks of the RND had been decimated by the fighting. Freyberg himself was badly wounded twice and had to be evacuated for hospital treatment. When the inevitable decision was made to evacuate the troops at Gallipoli, Freyberg was one of the last defenders to leave in early January 1916. He was also one of the most senior officers of his battalion (Hood Battalion) still alive. Among the dead was one of his brothers. The failure of the Gallipoli landings led to Churchill's resignation as First Lord of the Admiralty.

Freyberg and the remaining survivors of the RND returned to England in March 1916 to recuperate from their ordeal. With the outlook for the future role of the RND being uncertain, he sought, and succeeded, in transferring to the British Army as a captain. However, he remained seconded to the RND. The references from his previous commanding officers stressed his fearlessness and leadership qualities. While training in England, his connections allowed him to mix with many important people, including the Prime Minister, Mr Asquith, whose son served with him in Gallipoli.

Freyberg and the RND were next posted to France in May 1916, where they were retrained for trench warfare. By September, they were ready to take part in the next major offensive, the Battle of the Somme. This turned out to be one of the bloodiest of the First World War. The attack launched by the British forces was designed to help relieve pressure on the French troops fighting at Verdun. It began on July 1, preceded by a week of massive artillery bombardment that was expected to destroy most of the German defences, cut the barbed wire and allow the allied forces to march slowly forward in line with little resistance. The artillery bombardment, though very heavy, did not achieve its aims. When the allied infantry marched forwards as instructed, they found the barbed wire and the German army still more or less intact. They encountered a murderous fire, especially from well-protected German machine guns. That first day of the offensive was the blackest in British military history, with over 60,000 casualties, one-third of them killed. The battle raged continuously until November 18. The pattern was for one side to march across no-man's-land between the trenches to try to gain some territory. Usually this failed, with battalions being decimated in the process. If some territory was gained, the other side would then counterattack to regain the territory, again with terrible loss of life. Overall, it ended in stalemate, a battle of attrition. Fought over a relatively small front and with such ferocity and continuous artillery bombardment, the fighting and living conditions were unimaginable. Medical facilities were primitive and more troops died as a result of infected wounds than were killed outright by bullets and shrapnel.

Towards the very end of the battle, on November 13, Freyberg and the Hood battalion of the RND were involved in a major assault on the German trenches just a few hundred metres ahead of them in the region of the village of Beaucourt. Timed to commence in the early morning when it was still dark enough to provide some cover, the offensive was preceded by the usual heavy bombardment in an attempt to soften up the German trenches. During the course of the attack, whilst sustaining the expected high casualties, only the troops in Freyberg's area achieved their initial objective. In the resulting chaos, Freyberg gathered together the remnants of other isolated groups and continued his advance, capturing many German prisoners. Despite being seriously wounded, his success rallied neighbouring positions and resulted in a limited, but rare, success on the Somme. Freyberg was hospitalised and returned to

England. His exploits on the Somme were recognised with the award of the Victoria Cross. His heroic action can be more appreciated by the citation that read:

> *For most conspicuous bravery and brilliant leading as a Battalion Commander.*
>
> *By his splendid personal gallantry he carried the initial attack straight through the enemy's front system of trenches. Owing to mist and heavy fire of all descriptions, Lieutenant-Colonel Freyberg's command was much disorganised after the capture of his first objective. He personally rallied and re-formed his men, including men from other units who had become intermixed.*
>
> *He inspired all with his own contempt of danger. At the appointed time he led his men to the successful assault of the second objective, many prisoners being captured.*
>
> *During this advance he was twice wounded. He again rallied and re-formed all who were with him, and although unsupported in a very advanced position, he held his ground for the remainder of the day, and throughout the night, under heavy artillery and machine-gun fire. When reinforced on the following morning, he organised the attack on a strongly fortified village and showed a fine example of dash in personally leading the assault, capturing the village and five hundred prisoners. In this operation he was again wounded.*
>
> *Later in the afternoon, he was again wounded severely, but refused to leave the line till he had issued final instructions.*
>
> *The personality, valour and utter contempt of danger on the part of this single officer enabled the lodgement in the most advanced objective of the Corps to be permanently held, and on this* point d'appui *the line was eventually formed.*

Following his recuperation, Freyberg was back in the trenches of the Somme in February 1917. The winter conditions were very difficult, with more men again dying of cold and disease than from bullets. Towards the end of April, Freyberg left the Hood Battalion to take command of the 173 Infantry Brigade (a brigade consisting of up to 8000 troops). Now 28, he was the youngest (temporary) brigadier-general in the British Army. He was given the task of attacking the well-fortified German defensive position known as the Hindenburg Line. In one action in June, he disagreed strongly with the plan of attack but, after voicing his views, had little choice but to carry it out. His troops were repulsed with heavy loss of life.

After surviving action on the Somme, Freyberg and his troops were then posted to the battlefront at Ypres, known as Passchendale. Notorious for the mud, conditions here were probably the worst of the whole war. On 19 September 1917, Freyberg was again severely wounded and repatriated to England, where there were fears that he might not survive. However, he slowly recovered and by January 1918 was well enough to return to the mud of the Ypres-Passchendale sector and was given command of 88 Infantry Brigade.

At the beginning of 1918, the war on the Western Front had reached a critical situation. Following the stalemate of the last 3 years, the capitulation of Russia following the revolution meant that the Germans could reinforce their positions on the Western Front. A German offensive was launched on March 21, before the

Americans, newly entered into the war, could make their influence felt on behalf of the allies. Initially, the Germans made significant breakthroughs. Freyberg's brigade was moved along the line to help in a desperate rearguard action to blunt the progress of the offensive and allow for the controlled withdrawal of some of the allied forces. Both sides fought to exhaustion, and the German offensive petered out towards the end of April, with the usual heavy loss of life on both sides.

Fighting continued without much change of frontline positions until the beginning of August when the long war of attrition finally favoured the allies and the German front started to collapse. From this date, German forces were rapidly driven back. Freyberg, as always, was in the thick of the fighting. He received a first bar to his DSO for his actions around the town of Gheluvelt, the citation mentioning 'Wherever the fighting was hardest he was always to be found encouraging and directing his troops'.

With an armistice due to come into force on the eleventh hour on the eleventh day of the eleventh month of 1918, on the very last day of the war Freyberg was ordered yet again to risk his life by preventing a bridge from being blown up prior to the armistice. He carried out his task with flair and bravery and was awarded a second bar to his DSO.

In looking back on Freyberg's First World War experiences, his survival can only be regarded as miraculous. Here we have someone who was involved from the very beginning and was fighting right up to the very last minute. Against all odds, he survived the whole 4 years. He was always there leading his men from the front and was involved in all of the most bloody campaigns, namely Gallipoli, the Somme and Passchendale. More amazingly, he was seriously wounded on a number of occasions and, by rights, should have died in a field hospital, this being the time well before the discovery of antibiotics. He ended the war as one of the most decorated soldiers, holding the VC, the DSO with two bars, the CMG (Companion of the Most Distinguished Order of St Michael & St George) and was mentioned in despatches on six occasions.

Following the armistice, it would have been nice to report that Freyberg, realising his extreme luck in surviving the war, had returned to the peace and tranquillity of dental practise in England or New Zealand. However, sadly for the dental profession, this was not the case. He turned down a career in dentistry to continue in the British armed forces. By the time the Second World War started in 1939, he had risen to become general officer commanding all New Zealand forces.

The early days of the Second World War saw one defeat after another for Britain. With the fall of France and the evacuation of the British forces from Dunkirk, things were very bleak indeed. In England, Freyberg established his base in Colchester. He undertook the complex task of organising, training and staffing his forces. Also politically, he had to determine with the New Zealand government as to his powers of action on their behalf and how he was to interact with the generals of the British Army. He was now in a position to use his previous experience in the First World War. Then, when he disagreed with certain British commanders about what he regarded as unacceptable risks to men under his command, there was little he could do about it. Now he could influence policy. For the next 5 years

Figure 15.7 Lieutenant General Sir Bernard Freyberg at the battle of Cassino, January–May, 1944.
Source: This is photograph no. NA 10630 from the Imperial War Museum collection. http://en.wikipedia.org/wiki/File:Bernard_Freyberg.jpg. This artistic work created by the United Kingdom Government is in the public domain.

he was to deal at the highest level with all the leading personalities of different governments and armed forces. The fact that the New Zealand forces would have an exemplary record for conduct and service was a tribute to their training and his leadership. Out of a population of only 1,600,000, New Zealand eventually provided 135,000 citizens who served in the armed forces, of whom 11,000 died – as high a commitment as any other allied nation.

In the Second World War, Freyberg's exploits were even more adventurous and remarkable when compared with those from his First World War record, as he was now a general in charge of a large body of men and responsible for making important decisions in combat that affected thousands. Freyberg was involved in some of the most famous battles in the Second World War. The early phase of the war saw stunning victories by the Germans. Freyberg and his troops were forced into two initial retreats. He had to organise the retreat from Greece to Crete. He then was also forced to retreat from Crete after it fell to a parachute invasion by the Germans. Much has been written about the fall of Crete and whether Freyberg could be held partly responsible due to poor leadership and decision making. Despite the German victory at Crete, however, so many German paratroops were killed that such a method of invasion was never repeated.

When the tide of battle turned, Freyberg's army played a significant role in the battles of El Alamein and Monte Cassino (Figure 15.7). In reviewing and paying tribute to the victorious New Zealand forces in Tripoli in March 1943, British Prime Minister Winston Churchill referred to Freyberg as the 'salamander of the British Empire'. It is assumed that this was a reference to toughness and the ability of this amphibian to regenerate lost or damaged parts. Subsequently, the salamander appeared on Freyberg's coat of arms when he was later awarded his baronetcy.

Figure 15.8 Tombstone of Bernard Freyberg at St Martha's on the Hill, Surrey.

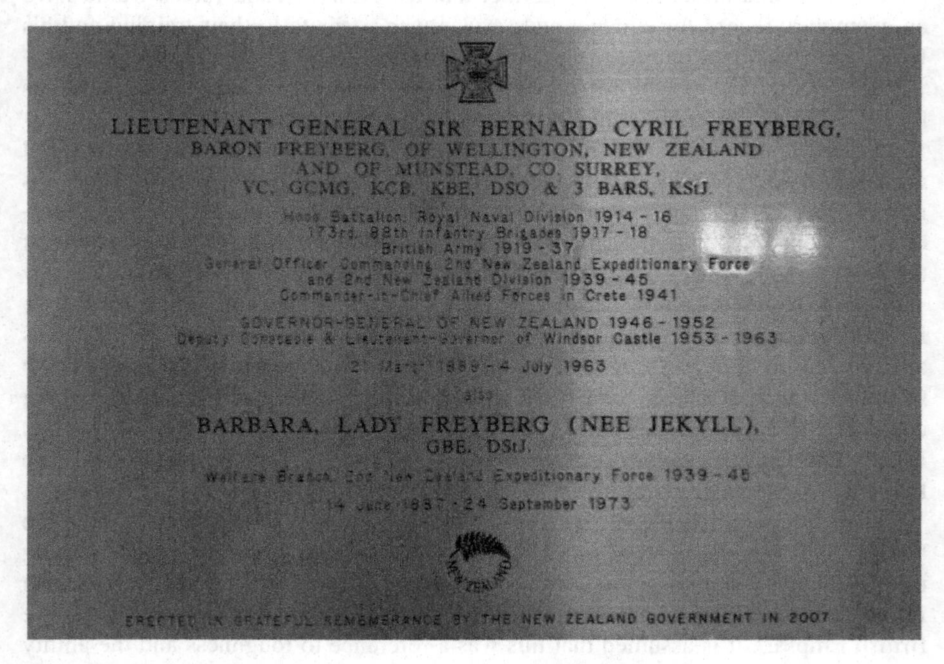

Figure 15.9 Commemorative plaque inside the church of St Martha's on the Hill, Surrey listing Bernard Freyberg and his wife's honours and offices.

In 1943 he received from King George VI the awards of the KCB (Knight Commander of the Most Honorable Order of the Bath) and the KBE (Knight Commander of the Most Excellent Order of the British Empire) for recognition of his outstanding wartime achievements.

After the capture of Rome, the rest of the year and the early months of 1945 saw Freyberg engaged in pursuit of the retreating German forces northwards, through Florence and then along the Adriatic Coast, through Padua and Venice, entering Trieste in May 1945. It was for his actions at Trieste, at the very end of Second World War, that Freyberg received a third bar to his DSO. During this campaign, he was wounded yet again, although this time it was due to damage sustained by his plane on landing badly.

After surviving both world wars and being at the centre of fighting throughout, Freyberg's services to Britain were not finished. At the age of 57, he was appointed governor-general of New Zealand from 1946 to 1952. He returned to England, where he was elevated to the Peerage. Baron Freyberg died on 5 July 1963, at the age of 74. The cause of death was a ruptured aorta that may have been related to the opening up of a wound received nearly 50 years earlier at Gallipoli.

Freyberg is buried in the grounds of a picturesque church, St Martha's, on the Hill, Surrey. His tombstone is simple and weathered (Figure 15.8). On it is engraved 'BERNARD FREYBERG, VC, 1889–1963, AN UNCONQUERABLE HEART'. Inside the church is a commemorative plaque listing all his principal honours and offices (Figure 15.9). For me, the most important detail missing is the word 'dentist'.

16 A Winning Smile: Beauty is in the Eye of the Beholder

For thousands of years people have been altering the appearance of their skin by using makeup, tattooing or deliberate scarring. They have even modified the shape of the skeleton by binding the feet or skull during childhood (Figure 16.1). As the face has a major function in communication by conveying feelings and emotions, it is not surprising that teeth feature prominently in intentional bodily modification.

The Smile

Which of the illustrations in Figure 16.2 do you find the most pleasing: A, B or C? Most people in Europe, Asia and America would probably choose Figure 16.2B. They would make Figure 16.2A their second choice as, compared with Figure 16.2B, the teeth are less regular and discoloured. Figure 16.2C would be a distant third. A person from certain parts of Africa might be unmoved by Figure 16.2A or B but swoon over the appearance of Figure 16.2C. Obviously, beauty is in the eye of the beholder, and what appeals to one person does not hold for another.

Through the widespread influence of the media, certain stereotypes have been promoted as icons of beauty. The public can be influenced to desire what they are told is the latest fashion. The appearance of the upper front teeth is particularly important as they are always on show. Having a 'nice' smile can boost one's confidence and self-esteem and help in career and marriage prospects. Much work has gone into deciding what factors constitute a desirable smile in the Western world. The angulation of the teeth, their size, shape, colour and proportions, and how they contact (Figures 16.3 and 16.4) are all part of the equation. Straight, white teeth with no spacing and with a perfect symmetry are general objectives.

As few people naturally have what is considered a perfect smile, many opt to have modification of their front teeth for cosmetic rather than clinical reasons. A popular procedure is to have false crowns or veneers placed over their filed-down front teeth. Despite a certain amount of discomfort, these procedures are relatively pain free (if not cost free!). Referring back to Figure 16.2, Figure 16.2A and B represents the same patient before and after cosmetic treatment.

In children whose teeth are crowded and the front ones unlikely to produce an attractive smile, at about ages 10–12 it is possible to take corrective measures by means of orthodontic treatment. Teeth farther back are extracted to provide the

Nothing but the Tooth. DOI: http://dx.doi.org/10.1016/B978-0-12-397190-6.00016-X

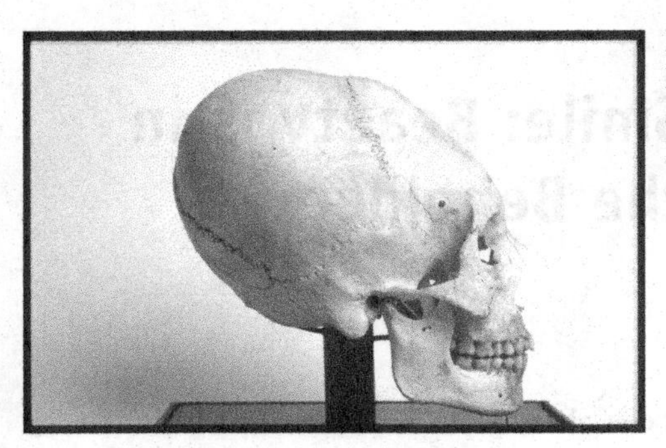

Figure 16.1 Egyptian skull whose shape has been altered by binding.
Source: Courtesy of the Hunterian Museum at the Royal College of Surgeons.

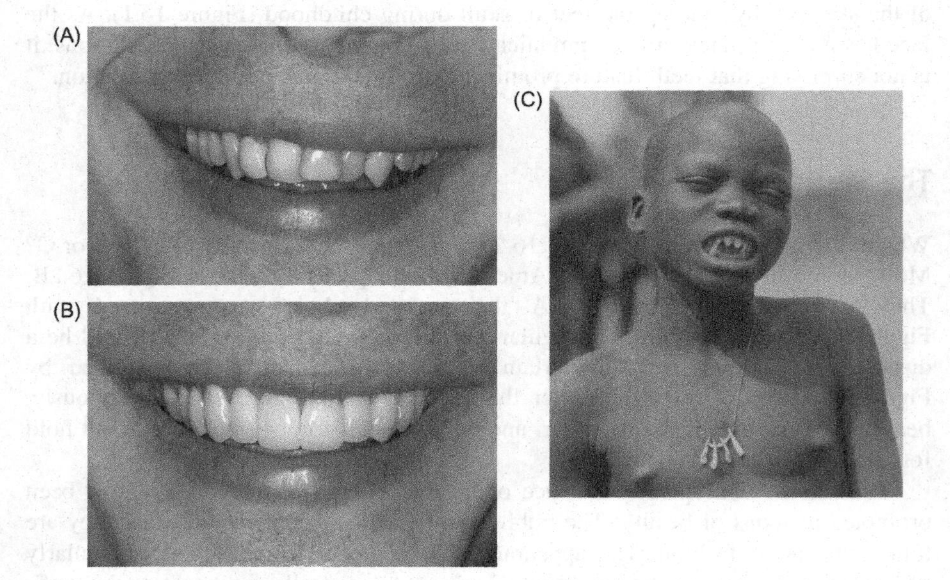

Figure 16.2 Three smiles. Figures 2A and B are white European. Figure 2C is a Moro girl from Southern Sudan. Her upper four incisors have been filed into points while her lower incisors have been extracted and strung with beads as a neck ornament.
Source: (A and B) Courtesy of Dr C. Orr. (C) Courtesy of Pitt Rivers Museum, University of Oxford, accession number PRM 1998.353.23.2.

necessary space to move and straighten those in front by a combination of springs and wires.

Some people choose to cap a front tooth with gold, such as Mike Tyson, the former world heavyweight boxing champion. One or two show business celebrities have had diamond inlays put in their front teeth. However, the ultimate example of this practise is a removable appliance known as a dental grille, made of gold and diamonds, that can be slipped over the front teeth whenever desired. Even as far back as the seventh century BC, female Etruscan skulls from Italy have been found

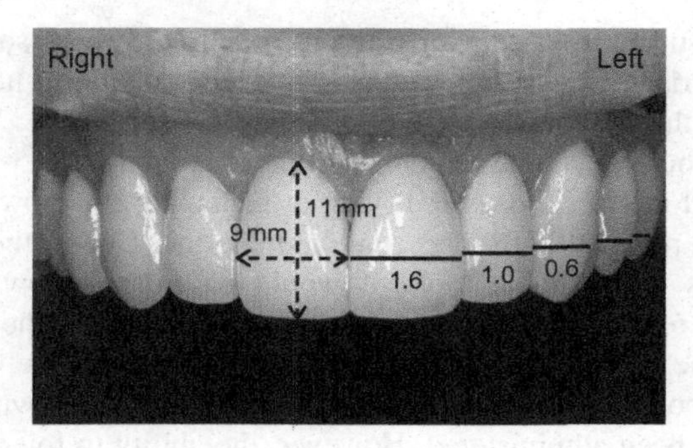

Figure 16.3 View of upper front teeth superimposed on which are desirable measurements for an ideal smile. For the right upper central incisor, ideal crown length is 11 mm and crown width is 9 mm, giving a width:crown ratio of 0.7–0.8. On the left side of the patient, the width proportions of the teeth are compared.
Source: Courtesy of Dr C. Orr.

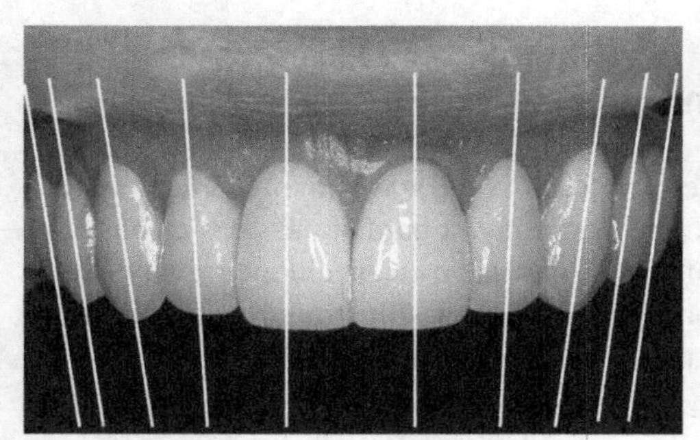

Figure 16.4 Subtle variations in the alignments of the long axis of the crowns of the maxillary teeth to help produce a perfect smile.
Source: Courtesy of Dr C. Orr.

with gold bands supporting false teeth. Dental modifications using precious metals were likely carried out to indicate the wealth and status of the owner and to enhance their attractiveness.

Cultural Significance

Over the last few thousand years and up to recent times, the inhabitants of Africa, North, Central (especially Mexico, Costa Rica, Honduras) and Southern America, South East Asia and Japan have incorporated dental modifications as an important part of their culture. Deliberately altering the appearance of normal healthy teeth is limited to the front teeth, especially the upper central incisors. This can take many forms, but the design may be so specific as to indicate the place of origin.

There are many reasons for the practise of dental modification. It may allow a cultural group a feeling of common identity and recognition; reflect the status of an individual, particularly where the techniques are lengthy and costly; be undertaken for religious reasons (as it can be identified in artefacts representing deities); be related to magic/superstition or with warding off disease; be associated with coming-of-age or marriage rituals; be used as a type of branding to help identify criminals; and help with the pronunciation of a language.

Young adults, both male and female, starting at around age 15–20, are the main subjects for dental modification. Modifying or reshaping the crown of a permanent tooth requires the removal both of enamel and dentine. Due to the absence of efficient rotary drills and suitable tools, it would have been slow (in view of the extreme hardness of the outer enamel) and painful (because of the sensitivity of the underlying dentine). Although pain-preventing substances were potentially available (such as cocoa leaves containing cocaine), it is not known whether any of these would have been administered. However, the ability to tolerate pain could be part of a coming-of-age ceremony. Dental modification could have been carried out by the individuals themselves. Alternatively, specialised practitioners may have been involved as they seemed to have been sufficiently skilled in their art to avoid exposing the central pulp (and causing death of the tooth with abscess formation), as this condition is rarely found in teeth that have been modified.

Types of Dental Modification

The technique for modifying the shape of a tooth could have involved first softening the surface with a special plant extract. For filing, a hard stone such as obsidian would have been employed. Pieces of tooth may have been chipped away rather than continuously filed. A kit used for this procedure is illustrated in Figure 16.5. Well over 50 different types of dental modifications are known. They can be classified as follows:

1. The tooth is filed at the sides to produce a point (Figures 16.6 and 16.7).
2. The cutting edge of the tooth is partly removed or indented (Figures 16.7 and 16.8).
3. One or both angles of the crown are removed (Figure 16.7).
4. The front surface of the tooth is scratched or grooved in various patterns (Figure 16.9).
5. One or more circular wells are cut into the front surface and a semi-precious stone placed into each (Figure 16.10).
6. Multiple combinations of the above.
7. Tooth extraction (ablation) (Figure 16.2C).

Where a portion of the sound tooth is cut away or the tooth extracted, the pink, fleshy tongue would be clearly visible behind, possibly enhancing the beauty of the smile. If the individual also chewed betel nut, this would colour the mouth and teeth a brownish red, adding to the perceived general attractiveness. Teeth were also blackened using derivatives from plants (Figure 16.6) such as the Chinese fever vine and flowers such as *Solanum incanum* (a species of Nightshade found in Africa containing nicotine).

In burials from the Mariana Islands off the Philippines, lines were scratched on the front surfaces of the teeth. Three main patterns could be distinguished: vertical lines, diagonal lines and cross-hatching (Figure 16.9). The number of lines usually

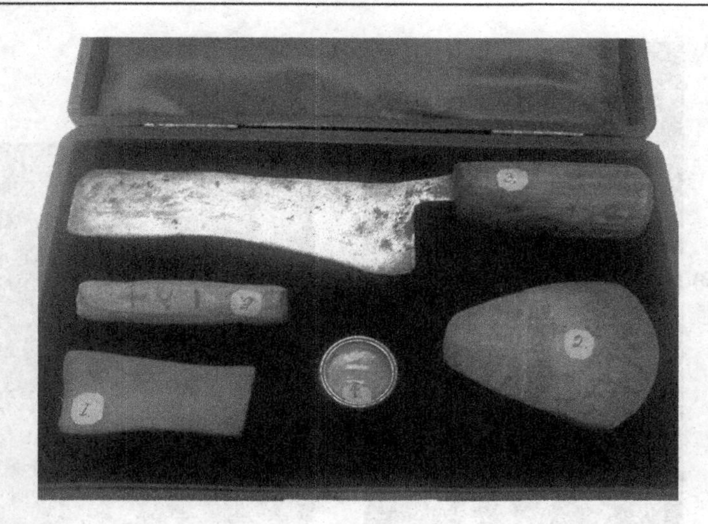

Figure 16.5 Kit used in Java for the ritual removal of the occlusal portions of the maxillary incisors and canines of boys in the late nineteenth century.
1. Smooth block of grey stone (possibly granite), presumably for filing the teeth.
2. Grooved shell (possibly coconut), presumably for filing.
3. Large saw-edged knife.
4. Two enamel flakes.
5. Cylindrical splint of wood; possibly used for patient to bite on while chipping/filing is performed.

Source: Courtesy of the Hunterian Museum at the Royal College of Surgeons.

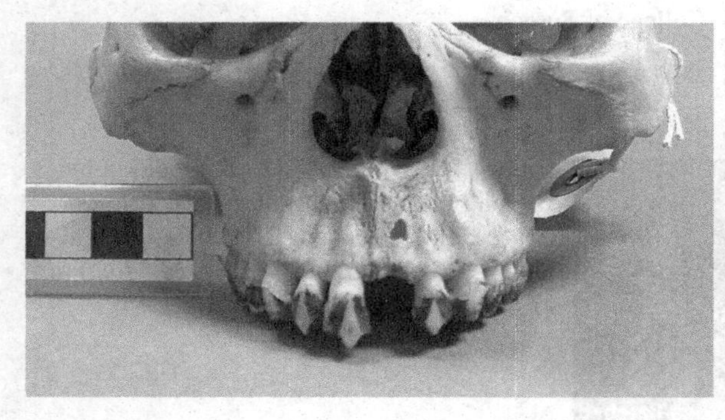

Figure 16.6 Dentally modified tooth from Africa. Three incisor teeth have been filed to a point and also stained black.
Source: Courtesy of the Hunterian Museum at the Royal College of Surgeons.

varied from about two to six, although one specimen had eight. The teeth were frequently stained reddish brown due to betel nut chewing. Prior softening of the enamel with plant extracts may have made it easier to incise the surface with hard flakes of rock such as chert or basalt. As this type of dental modification was relatively uncommon, it is more likely to represent status or lineage rather than a general rite of passage.

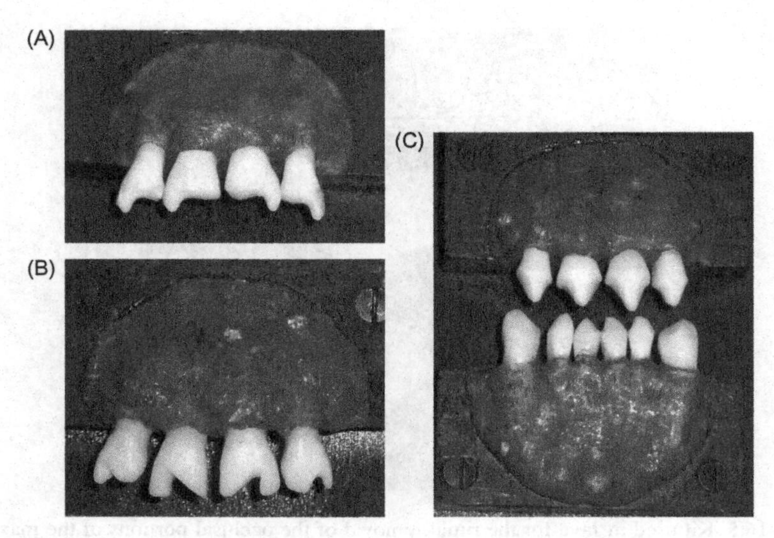

Figure 16.7 Part of a display showing three types of dental modifications from Central East Africa. (A) Labelled 'paired unilateral notch' from Mozambique. (B) Labelled 'diagonal notch' from Nyasaland. (C) Labelled 'shouldered peg' from Quilimane.
Source: Courtesy of Pitt Rivers Museum, University of Oxford, accession number PRM 1911.20.1-14. Teeth presented in 1911.

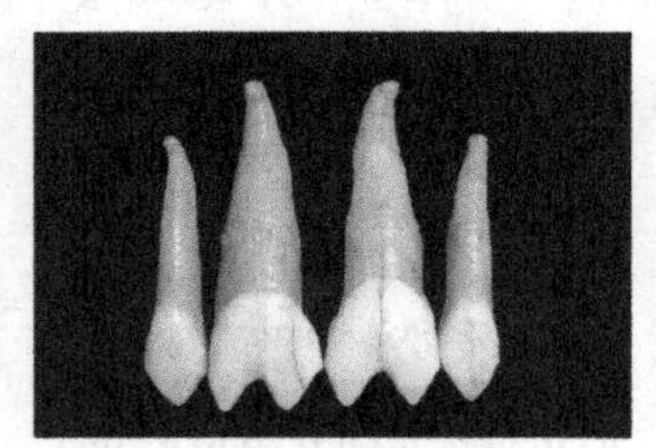

Figure 16.8 Modified upper four incisors from Africa. The incisal edges of the larger central incisors (middle two teeth) have been indented, while the lateral incisors at the sides are pointed.
Source: Courtesy of the Hunterian Museum at the Royal College of Surgeons.

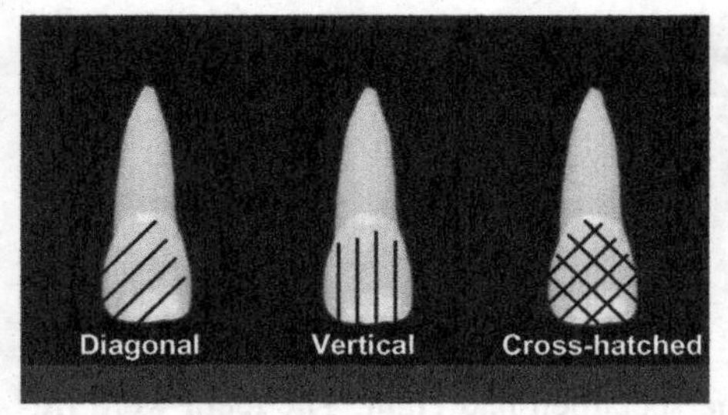

Figure 16.9 Patterns in teeth modified by incising lines on the front surface.
Source: Redrawn from R. Ikehara-Quebral and M.T. Douglas, 1997. *American Journal of Physical Anthropology*.

Modified Teeth from Europe

Until very recently, no evidence of intentional dental modification had been found anywhere in Europe. However, a detailed examination of over 550 skeletons of

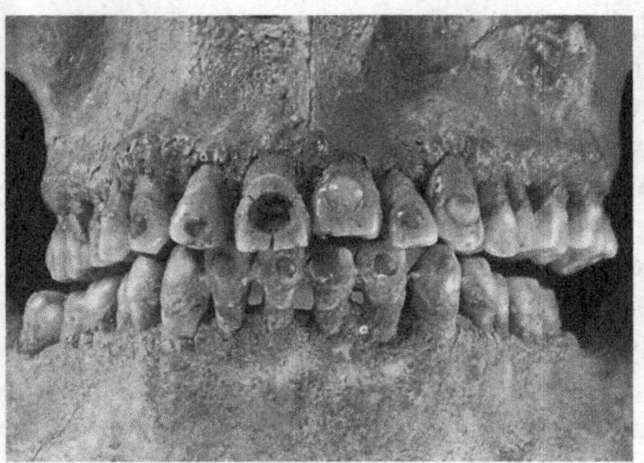

Figure 16.10 Mayan skull containing numerous semi-precious stones in upper and lower teeth. Seven inlays are of green jadeite, while two are of turquoise. Two inlays have not been retained, leaving empty cavities.
Source: Courtesy of the Director of Physical Anthropology, National Institute of Anthropology and History (INAH), Mexico.

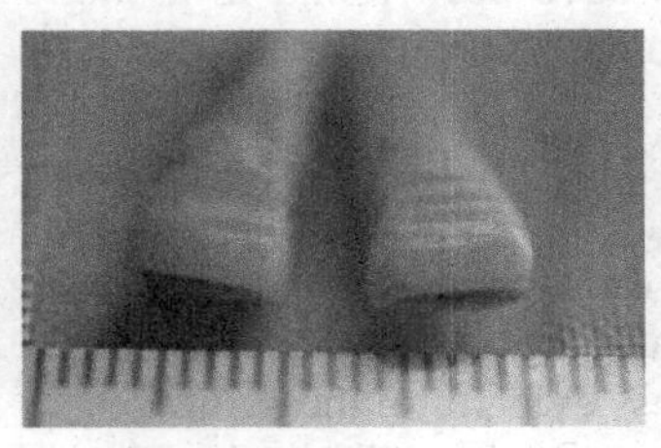

Figure 16.11 Two dentally modified upper incisor teeth from a Viking burial, in the form of fine horizontal grooves high up the front surface towards the gums. They would have been difficult to see in life.
Source: Courtesy of Oxford Archaeology and Dorset County Council.

men, women and children from four different Viking cemeteries in Southern Sweden did show evidence of it. The skeletons studied were from the period AD 800–1050. Twenty-four of the skulls, all males, most of them young, showed horizontal grooves in the upper front teeth. While a few had just one, most had two or three grooves per tooth, of varying depths. Often the grooves were present on quite a few of the upper front teeth of the same individual, the pattern varying on each tooth. Coincidentally, in the Viking burial of executed warriors referred to in Chapter 7 (page 97), one skull showed a similar type of dental modification to the Swedish ones on two of the upper front teeth (Figure 16.11).

The significance of these horizontally filed grooves in Viking men is not understood. Curiously, the grooves were placed high up near the gum rather than low down, making them barely visible unless the individual snarled to uncover the gums. Were they a mark of authority or trade? Were they a sign of distinction in battle, equivalent to awarding stars to a general? Could the teeth even have been modified after death? We may never know the answers to these questions.

Inlays of Semi-Precious Stones

Dental modification reached the height of sophistication with the Maya peoples of Mexico and Central America, who inserted inlays into the teeth. Among the semi-precious materials used were jade, turquoise, haematite and rock crystal. This procedure required very special skills. The first stage would have been to prepare a round cavity in the tooth, using a hard, tubular bit in combination with an abrasive paste. The bit, made of jade, copper or some other hard material, was manipulated by hand or by a bow drill. Once the cavity had been completed, a close-fitting inlay was inserted, sometimes held in place with the aid of an adhesive.

Jade comprises two distinct types of semi-precious stone of similar appearance: nephrite and jadeite. Chinese jade is principally nephrite; Mexican jade is principally jadeite and is mined in neighbouring Guatemala. It is the semi-translucent, intense green, even colour of the finest 'imperial' jadeite that makes the stone so prized. Like turquoise, it was fit only for the Gods and prepared by the finest craftsmen, as evidenced by the fact that over a thousand years later, inlays remain firmly in place (Figure 16.10). For this reason, jadeite used as a tooth inlay was available only to people of the highest status. Jadeite inlays can be found in upper and lower front teeth, and up to three jadeite inlays have been found in a single tooth. In some cases different semi-precious stones may be present in the teeth of the same individual.

Teeth containing inlays may also have had their shape modified, with different combinations in different teeth of the same individual. In addition to the pain suffered in having an inlay prepared and fitted, further excruciating pain would have followed in the rare instances where the operator made the cavity too deep, penetrating into the dental pulp and causing a dental abscess evident on the skull.

Hardly any dental modifications have been discovered in milk teeth. This is because milk teeth are small, rapidly worn away and lost early in life. Also the soft, sensitive, central dental pulp is proportionately large, while the covering, hard, tooth tissue is comparatively thin. This means that the dental pulp would be readily exposed were the tooth to be modified. However, one exquisite example in milk teeth does exist. The poorly preserved and partial remains of a child between age 4 and 5 were discovered in Belize, Central America, and dated at AD 700–800. Fifteen milk teeth were recovered, most having dropped out of the jaws. The two upper second incisors had both been dentally modified with one still containing a jadeite inlay. In addition, both teeth appeared to have had one of their edges (distal edge) filed. Bearing in mind the difficulties involved, it is believed the inlays were prepared after the death of the child, possibly indicating its high status.

Perhaps the ultimate example of dental modification is when one or more front teeth have been deliberately pulled out (ablation) (Figure 16.2C). This could have been for cultural reasons including beautification, coming of age, marriage or religion. Indeed, such a fashion fad has been observed among present-day youth of Cape Town, South Africa.

It has been suggested that in some instances tooth extraction was undertaken for linguistic reasons because a gap at the front of the mouth would result in a lisp when speaking and have allowed the individual to emphasise certain sounds important in the particular language. Although mainly undertaken in permanent teeth, there is evidence that milk teeth were ablated in babies, which could be associated with perceived health benefits.

Ritual tooth modification is little practised in the world today. However, one country where it can still be seen is Bali, where there is a Hindu coming-of-age

Figure 16.12 Photograph showing tooth dental modification in a Hindu coming-of-age ceremony known as Mepandes in Bali.
Source: Courtesy of Travelbild.com.

ceremony known as Mepandes. During this ceremony, the young person lies supine on a couch while a priest files the edges of the upper front six teeth (Figure 16.12). The reason for the filing rests in the belief that people are born with animal emotions and that filing their teeth will help keep control over the six deadly sins of lust, greed, anger, drunkenness, confusion and envy.

Tooth Necklaces

In addition to having their shape modified, teeth have played a prominent role as decorative objects worn by a person to signal their importance. Most commonly, this would be in the form of a necklace, which can be seen in civilisations throughout the world. Necklaces are made from the teeth of whatever animals are available and range from dolphins to kangaroos, antelope to fruit bats, bears to humans (see Figure 12.9). Tusks from boars can be placed through the nose of witch doctors and warriors and worn as armlets in Vanuatu (see Figure 2.17). If one's own necklace left a lot to be desired, all was not lost, as artificial animal and human teeth suitable for adding to necklaces were being exported from England to Africa around 1900.

Further Reading

With the advent of the computer and the internet, it is a simple task to gather information related to any topic in this book by typing the appropriate words in the search engine. However, to provide a guide to any reader wishing to delve further into the subject, some preliminary references are given. A number of references will, of necessity, relate to scientific journals, although these can be accessed from the British Library and in University Libraries. The particular journal websites will give free summaries and some may even provide the articles free.

Chapter 1

B.K.B. Berkovitz, R.P. Shellis, A longitudinal study of tooth succession in piranhas (Characidae) with an analysis of the tooth replacement cycle, Journal of Zoology, London 184 (1978) 545–561.

J.R. Quinn, Piranhas: fact and fiction, TFH Publications, New Jersey, 1992.

T. Roosevelt, Through the Brazilian wilderness, Wilder Publications, LLC, Virginia, 2008.

R.P. Shellis, B.K.B. Berkovitz, Observations on the dental anatomy of piranhas (Characidae) with special reference to tooth structure, Journal of Zoology, London 180 (1976) 69–84.

Chapter 2

S.M. Frank; Ingenious contrivances, curiously carved. Scrimshaw in the New Bedford Whaling Museum. New Bedford Whaling Museum, 2012.

J.T. Hagstrum, et al., Micrometeorite impacts in Beringian mammoth tusks and a bison skull, Journal of Siberian Federal University 3 (2010) 123–132.

W. Jackson, The use of unicorn horn in medicine, Pharmaceutical Journal 273 (2004) 925–927.

P. Lasco, M. Vickers, Ivory: a history and collectors guide, Thames and Hudson, London, 1987.

M. Locke, Structure of ivory, Journal of Morphology 269 (2008) 423–450.

A.E.W. Miles, J.W. White, Ivory, Proceedings of the Royal Society of Medicine 53 (1960) 775–780.

M.T. Nweeia, et al., Vestigial tooth anatomy and tusk nomenclature for *Monodon monocerus*, The Anatomical Record 295 (2012) 1006–1016.

M. Pederson, Gem and ornamental materials of organic origin, Elsevier Butterworth-Heinman, Oxford, 2004.

J. West, A.G. Credland, Scrimshaw: the art of the whaler, Hull City Museums and Art Galleries and Hutton Press, 1995.

Chapter 3

H.J. Bigelow, Insensibility during surgical operations produced by inhalation, Boston Medicine and Surgery Journal 35 (1846) 309–317.

F. Boott, Surgical operations performed during insensibility, The Lancet 2 January (1847) 5–8.

H. Davy, Researches, chemical and philosophical; chiefly concerning nitrous oxide or dephlogisticated nitrous air and its respiration, Johnson, London, 1800.

P.W. Ellsworth, On the modus operandi of medicines, Boston Medicine and Surgery Journal 32 (1845) 369–377.

F.D. Moore, John Collins Warren and his act of conscience: a brief narrative of the trial and triumph of a great surgeon, Annals of Surgery 229 (1999) 187–196.

L.F. Menczer, P.H. Jacobsohn, Dr Horace Wells: the discoverer of general anaesthesia, Journal of Oral and Maxillofacial Surgery 50 (1992) 506–509.

S.J. Snow, Blessed days of anaesthesia, Oxford University Press, Oxford, 2008.

J.C. Warren, Inhalation of ethereal vapour for the prevention of pain in surgical operations, Boston Medical and Surgical Journal 35 (1846) 376–379.

H. Wells, The discovery of the application of ether and other vapours to surgery, The Lancet 49 (1847) 471–474.

R.J. Wolfe, Tarnished idol: William Thomas Green and the introduction of surgical anaesthesia. A chronicle of the ether controversy, Norman Publishing, California, 2001.

R.J. Wolfe, L.F. Menczer (Eds.), I awaken to glory, Boston Medical Library, 1994.

Chapter 4

B.K.B. Berkovitz, The dentitions of amphibians and reptiles, Association for the Study of Reptilia and Amphibia – Monographs 1 (1) (1981).

C.C. Broomell, et al., Mineral minimization in nature's alternative teeth, Journal of the Royal Society Interface 4 (2007) 19–31.

F.P. Cuozzo, M.L. Sauther, Severe wear and tooth loss in wild ring-tailed lemurs (*Lemur catta*): a function of feeding ecology, dental structure, and individual life history, Journal of Human Evolution 51 (2006) 490–505.

M.L. Dalebout, D. Steele, C.S. Baker, Phylogeny of the beaked whale genus *Mesoplodon* (Ziphiidae: Cetacea) revealed by nuclear introns: implications for the evolution of male tusks, Systematic Biology 57 (2008) 857–875.

D. Desbruyeres, S. Hourdez, A new species of scale-worm (Polychaeta: Polynoidae), *Lepidonotopodium jouinae* sp. nov., from the Azores Triple junction on the Mid-Atlantic Ridge, Cahiers de Biologie Marine 41 (2000) 399–405.

S. Hillson, Teeth: Cambridge manuals in archaeology, Second edition, Cambridge University Press, 2005.

K. Jackson, The evolution of venom-conducting fangs: insights from developmental biology, Toxicon 49 (2007) 975–981.

P.W. Lucas, Dental functional morphology: how teeth work, Cambridge University Press, 2004.

A.E.W. Miles, C. Grigson, Colyer's variations and diseases of the teeth of animals, Cambridge University Press, 1990.

M.L. Sauther, R.W. Sussman, F. Cuozzo, Dental and general health in a population of wild ring-tailed lemurs: a life history approach, American Journal of Physical Anthropology 117 (2002) 122–132.

http://vertebratessi.edu/mammals/beaked_whales.This website contains much information on beaked whales and their teeth.

Chapter 5

S.S. Chavan, V.V. Yenni, Mandible like structure with fourteen teeth in a benign cystic teratoma, Indian Journal of Pathology and Medicine 52 (2009) 595–596.

M.C. Dean, C.F. Munro, Enamel growth and thickness in human teeth from ovarian teratomas (dermoid cysts). In: Current trends in dental morphology research. Thirteenth

International Symposium on Dental Morphology. Wydawnictwo Uniwersytetu Lodzkiego Lodz, Poland, 2005, pp. 371–382.
G. Falcinelli, et al., Modified osteo-odonto-keratoprosthesis for treatment of corneal blindness, Archives of Ophthalmology 123 (2005) 1319–1329.
A. Gomaa, O. Comyn, C. Liu, Keratosis in clinical practice – a review, Clinical and Experimental Ophthalmology 38 (2010) 211–224.
C.S.C. Liu, The eyes have it: a personal view, Book Guild Ltd, Brighton, 2012.
C.S.C. Liu, et al., Indications and technique of modern osteo-odonto-keratoprosthesis (OOKP), Eye News 4 (1998) 17–22.
C.S.C. Liu, et al., The osteo-odonto-keratoprosthesis (OOKP), Seminars in Ophthalmology 20 (2005) 113–128.
B. Strampelli, Keratoprosthesis with osteodental tissue, American Journal of Ophthalmology 89 (1963) 1029–1039.

Chapter 6

M.V. Chao, A conversation with Rita Levi-Montalcini, Annual Review of Physiology 72 (2010) 1–13.
W.M. Cowen, Viktor Hamburger and Rita Levi-Montalcini: the path to the discovery of nerve growth factor, Annual Review of Neuroscience 24 (2001) 551–600.
S. Cohen, Isolation of a mouse submaxillary gland protein accelerating incisor eruption and eyelid opening in the new-born animal, Journal of Biological Chemistry 237 (1962) 1555–1562.
S. Cohen, Origins of growth factors: NGF and EGF, Journal of Biological Chemistry 283 (2008) 33793–33797.
S. Cohen, R. Levi-Montalcini, A nerve growth-stimulating factor isolated from snake venom, Proceedings of the National Academy of Sciences of the United States of America 42 (1956) 571–574.
R. Levi-Montalcini, The nerve growth factor: thirty-five years later, Science 237 (1987) 1154–1164.
R. Levi-Montalcini, In praise of imperfection: my life and work, Basic Books, Inc., New York, NY, 1988.

Chapter 7

S.H. Ambrose, J. Buikstra, H.W. Krueger, Status and gender differences in diet at Mound 72, Cahokia, revealed by isotopic analysis of bone, Journal of Anthropological Archaeology 22 (2003) 217–226.
R. Amiot, et al., Oxygen isotopes from biogenic apatites suggest widespread endothermy in cretaceous dinosaurs, Earth and Planetary Science Letters 246 (2006) 41–54.
R. Amiot, et al., Oxygen isotope evidence for semi-aquatic habits among spinosaurid therapods, Geology 38 (2010) 139–142.
V.A. Andrushko, et al., Investigating a child sacrifice event from the Inca heartland, Journal of Archaeological Science 38 (2011) 323–333.
T.E. Cerling, et al., Diet of *Paranthropus boisei* in the early Pleistocene of East Africa, Proceedings of the National Academy of Sciences of the United States of America 108 (2011) 9337–9341.
M.T. Clementz, P.A. Holyroyd, P.L. Koch, Identifying aquatic habits of herbivorous mammals through stable isotope analysis, Palaios 23 (2008) 574–585.
M.T. Clementz, et al., Isotopic records from early whales and sea cows: contrasting patterns of ecological transition, Journal of Vertebrate Paleontology 26 (2006) 355–370.

R.A. Eagle, et al., Body temperatures of modern and extinct vertebrates from $^{13}C-^{18}O$ bond abundances and bioapatite, Proceedings of the National Academy of Sciences of the United States of America 107 (2010) 10377–10382.
R.A. Eagle, et al., Dinosaur body temperatures determined from isotopic ($^{13}C-^{18}O$) ordering in fossil biominerals, Science 333 (6041) (2011) 443–445.
B. Finucane, P.M. Agurto, W.H. Isbell, Human and animal diet at Conchopata, Peru: stable isotope evidence for maize agriculture and animal management practices during the middle horizon, Journal of Archaeological Science 33 (2006) 1766–1776.
A.J. Fitzpatrick, 2011. The Amesbury archer and the Boscombe Bowmen: v. 1: early Bell Beaker Burials at Boscombe Down, Amesbury, Wiltshire, Great Britain: excavations at Boscombe Down. Wessex Archaeology Reports, 2011.
H.C. Fricke, J. Hencecroth, M.E. Hoerner, Lowland–upland migration of sauropod dinosaurs during the Late Jurassic epoch, Nature 480 (2011) 513–515.
J.E.M. Hedges, R.E. Stevens, P.L. Koch, Isotopes in bones and teeth, Developments in Paleoenvironmental Research 10 (2006) 117–145.
A.G. Henry, et al., The diet of *Australopithecus sediba*, Nature (2012) Nature DOI: 10.1038/nature11185.
K.A. Hoppe, P.L. Koch, S.D. Webb, Tracking mammoths and mastodons: reconstruction of migratory behaviour using strontium isotope ratios, Geology 27 (1999) 439–442.
T.D. Price, V. Tiesler, J.H. Burton, Early African Diaspora in colonial Campeche, Mexico, American Journal of Physical Anthropology 130 (2006) 485–490.
M.P. Richards, E. Trinkaus, Isotopic evidence for the diets of European Neanderthals and early modern humans, Proceedings of the National Academy of Sciences of the United States of America 106 (2009) 16034–16039.
H. Schroeder, et al., Trans-Atlantic slavery: isotopic evidence for forced migration to Barbados, American Journal of Physical Anthropology 139 (2009) 547–557.
J. Staller, R. Tykot, B. Benz (Eds.), History of maize, Elsevier, London, 2006.
C. Stringer, The origin of our species, Allen Lane, London, 2011.
K.T. Uno, et al., Late Miocene to Pliocene carbon isotope record of differential diet change among East African herbivores, Proceedings of the National Academy of Sciences of the United States of America 108 (2011) 6509–6514.
T.D. White, et al., *Ardipithecus ramidus* and the paleobiology of early hominids, Archaeometry 44 (2002) 117–135.
www.wessexarch.co.uk/projects/amesbury/archer. This website contains much information on the Amesbury archer.
www.wessexarch.co.uk/projects/wiltshire/boscombe/bowmen. This website contains much information on the Boscombe Bowmen.

Chapter 8

Y. Kakizawa, et al., The histological structure of the upper and lower jaw teeth in the gobiid fish, *Sicyopterus japonicus*, Journal of Nihon School of Dentistry 68 (1986) 626–632.
P. Keith, et al., Characterisation of post-larval to juvenile stages, metamorphosis and recruitment of an amphidromous goby, *Sicyopterus lagocephalus* (Pallas) (Teleostei: Gobiidae: Sicydiinae), Marine and Freshwater Research 59 (2008) 876–889.
K. Mochizuki, S. Fukui, Development and replacement of upper jaw teeth in gobiid fish (*Sicyopterus japonica*), Japanese Journal of Ichthyology 30 (1983) 27–36.
K. Mochuzuki, S. Fukui, S. Fulneh, Development and replacement of teeth on the jaws and pharynx in a gobiid fish *Sicydium plumieri* from Puerto Rico, with comments on resorption of upper jaw teeth, Natural History Research 1 (1991) 41–52.

K. Moriyama, et al., Morphological characteristics of upper jaw dentition in a gobiid fish (*Sicyopterus japonicus*): a micro-computed tomography study, Journal of Oral Biosciences 51 (2009) 81–90.

K. Moriyama, et al., Plate-like permanent dental laminae of upper jaw dentition in adult gobiid fish, *Sicyopterus japonicus*, Cell Tissue Research 340 (2010) 189–200.

H.L. Schoenfuss, T.A. Blanchard, Metamorphosis in the cranium of postlarval *Sicyopterus stimpsoni*, an endemic Hawaiian stream goby, Micronesica 30 (1997) 93–104.

Chapter 9

H. Bandali, et al., Egg tooth development in snakes, European Cells and Materials 14 (Suppl. 2) (2007) 134.

B.K.B. Berkovitz, Tooth replacement in non-mammalian vertebrates, in: M.F. Teaford, M. M. Smith, M.W.J. Ferguson (Eds.), Development, function and evolution of teeth, Cambridge University Press, 2000, pp. 186–200.

B.K.B. Berkovitz, G.R. Holland, B.J. Moxham, A colour atlas and textbook of oral anatomy, histology and embryology, Fourth edition, Harcourt, Edinburgh, 2009, pp. 365–374

G.R. De Beer, Caruncles and egg-teeth: some aspects of the concept of homology, Proceedings of the Linnean Society of London 161 (1949) 218–224.

A.G. Edmond, Dentition, in: C. Gans, A. d'A bellairs, T.S. Parsons (Eds.), Biology of the reptiles, Academic Press, New York, NY, 1969, pp. 117–200.

A. Huysseune, Developmental plasticity in the dentition of a heterodont polyphyodont fish species, in: M.F. Teaford, M.M. Smith, M.W.J. Ferguson (Eds.), Development, function and evolution of teeth, Cambridge University Press, 2000, pp. 231–241.

C.A. Luer, P.C. Blum, P.W. Gilbert, Rate of tooth replacement in the nurse shark (*Ginglymostoma cirratum*), Copeia 1 (1990) 182–191.

B. Westergaard, M.W.J. Ferguson, Development of the dentition in *Alligator mississippiensis*: upper jaw dental and craniofacial development in embryos, hatchlings, and young juveniles, with a comparison to lower jaw development, American Journal of Anatomy 187 (1990) 393–421.

Chapter 10

R. Adell, et al., A 15-year study of osseointegrated implants in the treatment of the edentulous jaw, International Journal of Oral Surgery 10 (1981) 387–416.

P.-I. Branemark, Osseointegration and its experimental background, Journal of Prosthetic Dentistry 50 (1983) 399–410.

P.-I. Branemark, The osseointegration book: from Calvaria to Calcaneus, Quintessence Publishing, Berlin, 2005.

P.-I. Branemark, et al., Intra-osseous anchorage of dental prostheses, Scandinavian Journal of Plastic Reconstructive Surgery 3 (1969) 81–100.

P.-I. Branemark, G.A. Zarb, T. Albrektsson (Eds.), Tissue-integrated prostheses: osseointegration in clinical dentistry, Quintessence Publishing, Chicago, 1985.

P.-I. Branemark, et al., Osseointegrated implants in the treatment of the edentulous jaws. Experience from a 10-year period, Scandinavian Journal of Plastic and Reconstructive Surgery 16 (Suppl. 16) (1977) 1–132.

G. MacFarlane, Howard Florey: making of a great scientist, Oxford University press, Oxford, 1979.

E. McClarence, Close to the edge: Branemark and the development of osseointegration, Quintessence Publishing, London, 2003.

Chapter 11

B.K.B. Berkovitz, G.R. Holland, B.J. Moxham, A colour atlas and textbook of Oral Anatomy, Histology and Embryology, Fourth edition, Harcourt, Edinburgh, 2009, pp. 299–310

Y. Chen, et al., Conservation of early odontogenic signalling pathways in Aves, Proceedings of the National Academy of Science of the United States of America 97 (2000) 10044–10049.

S.J. Gould, Hen's teeth and horse's toes, W.W. Norton & Co, New York, NY, 1983.

M.P. Harris, et al., The development of archosaurian first-generation teeth in a chicken mutant, Current Biology 16 (2006) 371–377.

T.A. Mitsiadis, J. Caton, M. Cobourne, Waking up the sleeping beauty: recovery of the ancestral bird odontogenic program, Journal of Experimental Zoology (Molecular and Developmental Evolution) 306B (2006) 227–233.

T.A. Mitsiadis, et al., Development of chick embryos after mouse neural crest transplantations, Proceedings of the National Academy of Science of the United States of America 100 (2003) 6541–6545.

P.T. Sharpe, C.S. Young, Test tube teeth, Scientific American 295 (2005) 34–41.

I. Thesleff, Developmental biology and building a tooth, Quintessence International (2003) 117–128.

Chapter 12

B.K.B. Berkovitz, C. Grigson, C. Dean, Caroline Crachami, the Sicilian dwarf (1815–1824): was she really nine years old at death? American Journal of Medical Genetics 76 (1998) 343–348.

J. Bondeson, Caroline Crachami, the Sicilian fairy: a case of bird-headed dwarfism, American Journal of Medical Genetics 44 (1992) 210–219.

W. Hunter, Anatomia uteri umani gravidi tabulis, Baskerville Press, 1774.

H. Mantel, The giant, O'Brien, Harper Collins, London, 2010.

W. Moore, The knife man, Bantam Press, London, 2005.

R.F. Sognnaes, F. Strom, The odontological identification of Adolf Hitler. Definitive documentation by X-rays, interrogations and autopsy findings, Acta Odontologica Scandinavica 31 (1973) 43–69.

D.J.M. Wright, John Hunter and venereal disease, Annals of the Royal College of surgeons of England 61 (1981) 198–202.

http://home.tiac.net/~cri_a/piltdown/piltdown.html. This is a website with much information concerning the Piltdown forgery.

Chapter 13

T.G. Bromage, M.C. Dean, Re-evaluation of the age at death of immature fossil hominids, Nature 317 (1985) 525–528.

M.C. Dean, Tooth microstructure tracks the pace of human life-history evolution, Proceedings of the Royal Society B: Biological Sciences 273 (2006) 2799–2802.

M.C. Dean, et al., Growth processes in teeth distinguish modern humans from *Homo erectus* and earlier hominins, Nature 414 (2001) 628–631.

M.C. Dean, B.H. Smith, Growth and development of the Nariokotome youth, KNM-WT 15000, in: F.E Grine, J.C. Fleagle, R.E. Leakey (Eds.), The first humans – origin and early evolution of the genus *Homo*, Springer, New York/Heidelberg, 2009, pp. 101–120.

M.C. Dean, V.S. Lucas, Dental and skeletal growth in early fossil hominins, Annals of Human Biology 36 (2009) 545–561.

M.C. Dean, Retrieving chronological age from dental remains of early fossil hominins to reconstruct human growth in the past, Philosophical Transactions of the Royal Society London B: Biological Sciences 365 (2010) 3397–3410.

S.J. Gould, The Piltdown conspiracy, Hen's teeth and horses toes, Norton, New York, 1984, pp. 201–240

R.S. Lacruz, F.R. Rozzi, T.G. Bromage, Variation in enamel development of South African fossil hominids, Journal of Human Evolution 51 (2006) 580–590.

S. Milken, The prehistory of the mind: a search for the origins of art, religion and science, Thames and Hudson, London, 1996.

M.P. Richards, et al., Isotopic evidence for the intensive use of marine foods by late upper paleolithic humans, Journal of Human Evolution 49 (2005) 390–394.

T.M. Smith, et al., Dental evidence for ontogenetic differences between modern humans and Neanderthals, Proceedings of the National Academy of Science of the United States of America 107 (2010) 20923–20928.

C. Stringer., The origin of our species, Allen Lane, London, 2011.

Chapter 14

T. Barnes, Doc Holliday's road to Tombstone, Xlibris Corporation, 2005.

T. Cullen, The mild murderer: the true story of the Dr Crippen case, Houghton Mifflin Company, Boston, 1977.

R.F. Foran, et al., The conviction of Dr Crippen: new forensic findings in a century-old murder, American Journal of Forensic Science 56 (2010) 233–240.

J. Guinn, The last gunfight, Robson Press, London, 2011.

G.L. Roberts, Doc Holliday: the life and legend, Wiley, New Jersey, 2006.

K.H. Tanner, Doc Holliday: a family portrait, University of Oklahoma Press, 1998.

K.D. Watson, Dr Crippen: crime archive, The National Archive, Her Majesty's Stationery Office, London, 2007.

www.oldbaileyonline.org/browse.jsp?id=t19101011-74&div=t19101011-74. This website contains the full trial transcript of Dr Crippen's murder trial.

www.drcrippen.co.uk. This website contains much information concerning the Dr Crippen case and trial.

www.drcrippen.co.uk/features/le_neve_trial_transcript.html. This website contains the full trial transcript of Ethel Le Neve murder trial.

F. Young, The trial of Hawley Harvey Crippen, Ed: with notes and an introduction (classic reprint), Forgotten Books, 2012.

Chapter 15

L. Baber, J. Tonkin-Covell, Freyberg: Churchill's salamander, Hutchinson, London, 1989.

D.H. Fischer, Paul Revere's ride, Oxford University Press, Oxford, 1994.

E. Forbes, Paul Revere and the world he lived in, Mariner Books, Boston, 1999.

P. Freyberg, Bernard Freyberg, VC. Soldier of two nations, Hodder & Stoughton, London, 1991.

M. Gladwell, The tipping point, Abacus, London, 2000.

P. Singleton-Gates, General Lord Freyberg. VC. An unofficial biography, Michael Joseph, London, 1963.

Henry Wadsworth Longfellow's poem *Paul Revere's Midnight Ride* can be found at the following website: http://poetry.eserver.org/paul-revere.html

Chapter 16

G.E. Brasewell, M.R. Pitcavage, The cultural modification of teeth by the ancient Maya: a unique example from Pusilha, Belize, Mexicon 31 (2009) 24–27.

S. Fastlicht, Tooth mutilations and dentistry in Mexico, Quintessence Publishing, Berlin, 1976.

S. Hillson, Dental anthropology, Cambridge University Press, New York, NY, 1996.

R. Ikehara-Quebral, M.T. Douglas, Cultural alteration of teeth in the Mariana Islands, American Journal of Physical Anthropology 104 (1997) 381–391.

C. Larsen, Dental modifications and tool use in the Western Great Basin, American Journal of Physical Anthropology 67 (1985) 393–402.

G. Milner, C. Larsen, Teeth as artifacts of human behavior: intentional mutilation and accidental modification in: M.A. Kelly, C.S. Larsen (Eds.), Advances in Dental Anthropology. Wiley-Liss, New York, 1991, pp 357–78.

J.P. Mower, Deliberate ante-mortem dental modification and its implication in archaeology, ethnography and anthropology, Papers from the Institute of Archaeology 10 (1999) 37–53.

J. Romero, Dental mutilation, trephination and cranial deformation. In: Physical anthropology. Handbook of Mid American Indians. Edited by Stewart, TD, 1970.

G. Scott, C. Turner II, The anthropology of modern human teeth, Cambridge University Press, New York, NY, 1997.

N. Taylor, Tooth ablation in prehistoric Southeast Asia, International Journal of Osteoarchaeology (1996) 333–345.

Printed in the United States
By Bookmasters